SpringerBriefs in Applied Sciences and Technology

More information about this series at http://www.springer.com/series/8884

Gianluca Coccia · Giovanni Di Nicola
Alejandro Hidalgo

Parabolic Trough Collector Prototypes for Low-Temperature Process Heat

 Springer

Gianluca Coccia
Dipartimento di Ingegneria Industriale e
  Scienze Matematiche
Università Politecnica delle Marche
Ancona
Italy

Alejandro Hidalgo
Universidad Carlos III de Madrid
Madrid
Spain

Giovanni Di Nicola
Dipartimento di Ingegneria Industriale e
  Scienze Matematiche
Università Politecnica delle Marche
Ancona
Italy

ISSN 2191-530X   ISSN 2191-5318   (electronic)
SpringerBriefs in Applied Sciences and Technology
ISBN 978-3-319-27082-1   ISBN 978-3-319-27084-5   (eBook)
DOI 10.1007/978-3-319-27084-5

Library of Congress Control Number: 2015960408

Printed on acid-free paper

This Springer imprint is published by SpringerNature
The registered company is Springer International Publishing AG Switzerland

Preface

This book is a short but concise summary of prototypes of parabolic trough solar collectors (PTCs) presented in the technical literature and used for low-enthalpy (low-temperature) demands. The idea of writing this book started from a paper that two of the Authors wrote last year, which dealt with the design and manufacture of a low-temperature and low-cost PTC prototype developed in the Department of Industrial Engineering and Mathematical Sciences of Marche Polytechnic University (UNIVPM). Since that work brought us to deepen the field of PTCs adopted for thermal industrial processes, in particular low-temperature prototypes usually developed in the academic world, we taught that it could be useful and interesting to provide a book concerning these fascinating solar collectors.

We believe that this work is particularly addressed to researchers who have fundamentals of solar thermal energy, theoretical and/or practical, and intend to enhance their knowledge of the field of PTCs starting from low-temperature prototypes. Students of Engineering faculties could also be interested in the topic of this book, if their purpose is to study more advanced applications of renewable energy and applied physics.

The book is divided into five chapters. The introduction is presented in Chap. 1, which presents some fundamentals of solar thermal energy, including the characteristics of the sun. In addition, a classification of solar collectors available in the market is given, and concentrating collectors, among which PTCs, are described. Solar thermal applications typically used nowadays are also discussed.

Chapter 2 is the longest chapter of the book, providing a detailed description of the physical/mathematical modeling of PTCs. We decided to organize this chapter into three sections. The first section is a discussion of the angles which we refer to in order to calculate the position of the sun. As concerns PTCs, these angles are important to determine the correct slope necessary to the tracking system of the collector and to calculate the angle of incidence, a quantity which has a central role in solar collectors, and even more in PTCs. The second section presents the optical analysis of a PTC, i.e., all the aspects regarding the geometry of a PTC and the selected materials that influence its optical efficiency. Finally, the third section deals with the thermal analysis of a PTC, in other words the energy balance of the receiver and, hence, the thermal efficiency of a PTC.

The standards used to assess the performance of solar thermal collectors such as PTCs are discussed in Chap. 3. The procedures required for testing PTCs will be discussed in detail, focusing the attention on the most relevant parameters that should be measured in a solar collector. Moreover, uncertainty in thermal efficiency testing will be described in detail, along with some details on quality test methods.

Chapter 4 presents an overview of the manufacture of several concentrators of PTC prototypes available in literature. PTC concentrators are used to reflect the solar radiation to the other element of a PTC which will be treated in the last chapter, the receiver. Having the shape of a cylindric parabola, PTC concentrators are able to reflect each normal incident solar beam to a line belonging to the parabola itself and called focal line, where the receiver is located. These systems require an accurate design and manufacture in order to correctly concentrate the solar radiation, and they should be built adopting materials with good mechanical and optical properties. All these aspects will be discussed in the chapter in object.

Receivers, the elements of a PTC where the solar radiation is concentrated and collected, are described in Chap. 5. In particular, this chapter presents an overview of the receivers used in PTC prototypes, providing also some characteristics about the adopted materials. PTC receivers have a tubular form and are placed in the focal line of the concentrator. A fraction of solar energy absorbed by the receiver is transferred to the heat transfer fluid which circulates inside the tube, obtaining a useful heat gain. Thus, a correct design of such components is crucial. Along with a description of available receivers, this chapter discusses the performance of the presented prototypes. Finally, nanofluids, novel heat transfer fluids which seem to have very interesting thermophysical properties, are introduced.

Ancona, Madrid   Gianluca Coccia
November 2015   Giovanni Di Nicola
                                                                         Alejandro Hidalgo

Acknowledgments

We would like to thank all the people who worked with us in our PTC project: the technicians of DIISM (Department of Industrial Engineering and Mathematical Sciences) and all the students with whom we had the pleasure to share our experience.

As concerns students, we would like to express our special gratitude to everyone who contributed to realize some sections of the present book. We decided not to report the entire list of the people involved in the project, as it would be extremely long; we hope that our decision will be shared by everyone.

However, we cannot forget to be grateful to Eng. Marco Sotte, Ph.D., who originally undertook at DIISM an experimental program called PTC project which marked the beginning of our work on parabolic trough collectors.

We hope that this book will be appreciated by all of them.

Contents

Nomenclature

Latin Symbols

A_a	Aperture area (m^2)
A_{ae}	Effective aperture area (m^2)
A_f	Ratio of ineffective area to the whole aperture area
A_i	Ineffective area due to end effects (m^2)
A_r	Receiver area (m^2)
a	Accommodation coefficient
B	Day of the year factor (°)
b	Interaction coefficient
C	Concentration ratio
C_{max}	Maximum concentration ratio
$C_{max,2D}$	Maximum concentration ratio for two-dimensional concentrators
c_p	Specific heat at constant pressure (J kg^{-1} K^{-1})
D	Diameter (m), minimum diameter of the receiver for perfect reflectors (m)
D_{ai}	Inner absorber diameter (m)
D_{ao}	Outer absorber diameter (m)
D_{ci}	Inner cover diameter (m)
D_{co}	Outer cover diameter (m)
d	Intersection of the bounding parabola with the rim of the trough (m)
d^*	Universal nonrandom error parameter due to receiver mislocation and reflector profile errors
$(d_r)_y$	Receiver mislocation along the optical axis (m)
E	Equation of time (min)
F'	Collector efficiency factor
F_{cyl}	Form factor for concentric cylinders
F_R	Heat removal factor
f	Focal length/distance (m), friction factor
G_b	Beam radiation on a horizontal surface (W m^{-2})

G_{bn}	Beam radiation on a normal surface (W m^{-2})
G_{bt}	Beam radiation on a tilted surface (W m^{-2})
h_{air}	Convective heat transfer coefficient of the air (W m^{-2} K^{-1})
h_{ann}	Convective heat transfer coefficient of the annulus gas (W m^{-2} K^{-1})
h_f	Convective heat transfer coefficient of the fluid (W m^{-2} K^{-1})
K	Extinction coefficient (m^{-1}), capacity factor
$K_{\tau\alpha}$	Incident angle modifier
k	Mean free path between collisions of a molecule (m)
L	Length of the receiver (m)
L_c	Length of the concentrator mirror (m)
L_{loc}	Longitude (°)
L_{st}	Standard meridian for the local time zone (°)
$\dot{m}$	Mass flow rate (kg s^{-1})
Nu_{air}	Nusselt number of the air
Nu_f	Nusselt number of the fluid
n	Day of the year, refractive index, number of reflections, normal
Pr_{air}	Prandtl number of the air
Pr_{ann}	Prandtl number of the annulus air
Pr_f	Prandtl number of the fluid
p	Pressure (bar), annulus gas pressure (mmHg)
$Q_{c,ac}$	Convection from the absorber to the cover (W)
$Q_{c,af}$	Convection from the absorber to the fluid (W)
$Q_{c,ce}$	Convection from the cover to the environment (W)
$Q_{k,a}$	Conduction through the absorber (W)
$Q_{k,c}$	Conduction through the cover (W)
Q_l	Heat loss from the absorber (W)
Q_r	Energy emitted by the receiver (W)
$Q_{r,ac}$	Radiation from the absorber to the cover (W)
$Q_{r,ce}$	Radiation from the cover to the environment (W)
Q_u	Useful heat gain of the fluid (W)
Ra_{air}	Rayleigh number of the air
Ra_{ann}	Rayleigh number of the annulus air/gas
Re_f	Reynolds number of the fluid
Re_{air}	Reynolds number of the air
r	Radius (m), specular reflectance, local mirror radius (m)
r_r	Mirror radius (m)
$r_\parallel$	Parallel component of the specular reflectance
$r_\perp$	Perpendicular component of the specular reflectance
S	Distance between the receiver rim and the concentrator rim (m), beam radiation collected by the absorber (W)
S_a	Solar energy intercepted by the collector aperture area (W)
S_c	Beam radiation reflected toward the receiver (W)
T_{ai}	Inner absorber temperature (°C)
T_{air}	Ambient/air temperature (°C)

$T_{air,m}$	Mean ambient temperature (°C)
T_{ao}	Outer absorber temperature (°C)
T_{ci}	Inner cover temperature (°C)
T_{co}	Outer cover temperature (°C)
T_{dp}	Dew point ambient temperature (°C)
T_{fi}	Inlet fluid temperature (°C)
T_{fm}	Mean fluid temperature (°C)
$T_{f,max}$	Stagnation temperature (°C)
T_{fo}	Outlet fluid temperature (°C)
$T_{fo,s}$	Stabilized outlet temperature (°C)
T_r	Temperature of the receiver surface (K), average temperature of the absorber surface (°C)
T_{sky}	Sky temperature (°C)
t	Time (s), thickness of the cover (m)
U_L	Overall loss coefficient (W m^{-2} K^{-1})
W_a	Aperture of the parabola (m)
W_c	Aperture of the concentrator mirror (m)
w	Aperture of the parabola (m)
w'	Aperture of the parabolic segment (m)
w_{air}	Wind velocity (m s^{-1})
w_f	Velocity of the fluid (m s^{-1})
x	Abscissa
y	Ordinate
z	Height

Greek Symbols

α	Absorptance of the absorber
α_c	Absorptance of the cover
α_s	Solar altitude angle (°)
β	Slope (°), angle between the central solar ray and the normal to concentrator aperture plane (°)
β^*	Universal nonrandom error due to angular errors (°)
γ	Surface azimuth angle (°), intercept factor, ratio of specific heats for the annulus gas
γ_s	Solar azimuth angle (°)
δ	Declination (°), molecular diameter of the annulus gas (m)
ε_a	Emissivity of the absorber
ε_c	Emissivity of the cover
ε_{sky}	Sky emissivity
η	Thermal efficiency
η_o	Optical efficiency
θ	Angle of incidence (°)
θ_m	Half-acceptance angle (°)

θ_z	Zenith angle ($^\circ$)
λ	Wavelength (m), conductive heat transfer coefficient (W m^{-1} K^{-1})
λ_a	Conductive heat transfer coefficient of the absorber (W m^{-1} K^{-1})
λ_{air}	Conductive heat transfer coefficient of the air (W m^{-1} K^{-1})
λ_{ann}	Conductive heat transfer coefficient of the annulus air (W m^{-1} K^{-1})
λ_{eff}	Effective conductive heat transfer coefficient of the annulus air (W m^{-1} K^{-1})
λ_f	Conductive heat transfer coefficient of the fluid (W m^{-1} K^{-1})
λ_{std}	Conductive heat transfer coefficient of the annulus gas at standard temperature and pressure (W m^{-1} K^{-1})
ν	Kinematic viscosity (m^2 s^{-1})
ξ	Pipe roughness (m)
π	Pi
ρ	Specular reflectance of the concentrator, density (kg m^{-3})
ρ_c	Diffuse reflectance of the cover
$\rho_\perp$	Perpendicular component of the diffuse reflectance
σ	Stefan–Boltzmann constant (W m^{-2} K^{-4})
σ^*	Universal random error parameter
$\sigma_{mirror,n}$	Standard deviation of the distribution of diffusivity of the reflective material at normal incidence
$\sigma_{slope,n}$	Standard deviation of the distribution of local slope errors at normal incidence
$\sigma_{tot,n}$	Total reflected energy distribution standard deviation at normal incidence
τ	Transmittance of the cover
τ_a	Transmittance of the cover by considering only absorption
τ_r	Transmittance of the cover by considering only reflection/refraction
$\tau_\perp$	Perpendicular component of the transmittance
$(\tau\alpha)$	Transmittance–absorptance product
ϕ	Latitude ($^\circ$), angle between the center line and a generic beam reflected at the focus ($^\circ$)
ϕ_r	Rim angle ($^\circ$)
ω	Hour angle ($^\circ$)

Acronyms

ANSI	American National Standards Institute
ASHRAE	American Society of Heating, Refrigerating, and Air-Conditioning Engineers
CVD	Chemical vapor deposition
DC	Direct current
GUM	Guide for Uncertainty in Measurements
HTF	Heat transfer fluid
ISO	International Organization for Standardization

LCT	Local clock time
MCNT	Multiwalled carbon nanotube
MDF	Medium density fiber
PCM	Phase change material
PTC	Parabolic trough collector
PVD	Physical vapor deposition
SEGS	Solar Electric Generating System
UV	Ultraviolet
XEPS	Extruded polyester

About the Authors

Gianluca Coccia is a Ph.D. student of Industrial Engineering at the Department of Industrial Engineering and Mathematical Sciences of Marche Polytechnic University, Italy. During his Bachelor and Master theses, he studied parabolic trough solar collectors working with two experimental prototypes and developing a mathematical model able to determine the optical and the thermal efficiency of such systems. His doctoral activity includes properties of fluids (thermal conductivity, dynamic viscosity, surface tension, virial coefficients), nanofluids and solar energy systems (parabolic trough collectors and solar cookers). He also studies mathematical models and artificial neural networks applied to the aforementioned topics. He is author of papers published in international journals in the field of solar energy and thermophysical properties of fluids.

Giovanni Di Nicola is associate professor in Environmental Applied Physics at the Faculty of Engineering of the Marche Polytechnic University, Italy. During the last 20 years, he conducted research, both experimental and theoretical, on thermophysical properties with particular attention to environmental friendly refrigerants. He also developed theoretical models for the prediction of surface tension, thermal conductivity, dynamic viscosity and virial coefficients of organic fluids. During the last five years, he turned his attention to the sector of solar thermal energy, in particular studying parabolic trough collectors (PTCs) and solar cookers. He is a member of the editorial board of the International Journal of Thermophysics, of the Italian Thermophysical Properties Association (AIPT) and of the Commission B1 (thermodynamic and transport properties) of the International Institute of Refrigeration (IIR). He is author of over 70 articles in international journals with expert reviewers.

Alejandro Hidalgo studied Industrial Technologies in the University Carlos III of Madrid. He started to work in the field of solar energy thanks to an Erasmus programme in the Marche Polytechnic University, Italy, under the supervision of Prof. Giovanni Di Nicola. During his experience, he worked in experimental projects involving parabolic trough solar collectors and solar cookers.

Chapter 1
Introduction

Abstract In this introductory chapter, fundamentals of solar thermal energy are discussed and some details concerning the sun, the star of our solar system and the source of solar energy on earth, are given. Further, solar thermal collectors available in the market are briefly presented, focusing in particular on concentrating collectors, systems able to concentrate and hence to increase the solar radiation. Among these, parabolic trough collectors (PTCs), which are the main subject of the present book, are described by dividing them into two different parts: the concentrator and the receiver. The materials used in such solar collector and their working principle are also discussed. Finally, typical solar thermal applications are presented, in particular as regards their utilization with PTCs.

Keywords Energy consumption · Sun · Solar collectors · Thermal applications · Industrial process heat

1.1 Fundamentals of Low-Temperature Solar Thermal Energy

The growth in energy consumption over the past 20 years has been significant and demand for energy will continue to grow due to global population increase. The global commercial low-temperature heat consumption is estimated to be about 10 EJ per year only for hot water production [1]. The industrial energy consumption in the industrialized countries accounts for 30% of the total required energy; in Europe, two-thirds of this energy consists of heat [2]. The only way to meet this global heat demand without contributing to climate change and environmental problems implies the utilization of renewable sources.

Solar energy is the most abundant permanent energy resource on Earth. One of the most popular low-temperature application of solar system is domestic water heating. Beyond the domestic applications, solar energy has several potential fields of application for low-temperature industrial processes. A wide range of collectors can be used for these low-temperature applications: flat plate, evacuated tube, compound

© The Author(s) 2016
G. Coccia et al., *Parabolic Trough Collector Prototypes for Low-Temperature Process Heat*, SpringerBriefs in Applied Sciences and Technology,
DOI 10.1007/978-3-319-27084-5_1

parabolic and more advanced types such as parabolic trough collectors (PTCs), which
appear to be one of the most promising technologies to use the energy of solar
radiation [3].

The source of the solar radiation is the Sun, the star at the center of the solar system.
It is almost perfectly spherical and consists of hot gaseous matter. The nature of
energy generation in the sun is still an unanswered question. Spectral measurements
have confirmed the presence of nearly all the known elements in it; however, 80%
of the Sun is hydrogen and 19% is helium [4].

Several hydrogen-to-helium thermonuclear fusion reactions are the source of the
solar energy. One of the most important is:

$$4_1{}^1\text{H} \longrightarrow {}^4_2\text{He} + 2e^+ + 24.56\,\text{MeV} \tag{1.1}$$

where e^+ indicates a positron, a particle having mass equal to that of an electron and
opposite electrical charge. The mass of the helium nucleus is lower than that of four
protons, mass having been lost in the reaction and converted to energy. Actually, the
Sun is a continuous fusion reactor with its constituent gases retained by enormous
gravitational forces.

The energy produced in the interior of the solar sphere at temperatures variously
estimated at $8\text{--}40\times10^6$ K must be transferred out to the surface and then be radiated
into space. A succession of radiative and convective processes occur with successive
emission, absorption and re-radiation. However, the nature of the energy-creation
process is of no importance to terrestrial users of the solar radiation. Of interest is
the amount of energy, its spectral and temporal distribution, and its variation with
time of day and year.

Solar energy is the world's most abundant permanent source of energy. The amount
of solar energy intercepted by the Earth is 5000 times greater than the sum of different
energies: terrestrial gravitational, nuclear and geothermal, and lunar gravitational [4].
Physical and orbital characteristics of the Sun of engineering interest are provided
in Table 1.1.

Table 1.1 Characteristics of the Sun of engineering interest

Characteristic	Value
Mean distance from Earth (m)	1.496×10^{11}
Mean diameter (m)	1.393×10^9
Surface area (m^2)	6.088×10^{18}
Mass (kg)	1.987×10^{30}
Power emission (W)	3.839×10^{26}
Power intercepted by Earth (W)	1.731×10^{17}
Blackbody temperature (K)	5777

1.2 Solar Thermal Collectors

The principle usually followed in solar thermal energy collection is to expose a dark surface to solar radiation so that this one is absorbed. A fraction of this absorbed radiation is then transferred to a heat transfer fluid (HTF) such as water or air.

Solar collectors can be divided into stationary (or non-concentrating) and concentrating systems. The main difference between them is that the formers have the same area for intercepting and absorbing solar radiation, while the latters generally have reflecting/refracting surfaces which intercept and focus the solar beam radiation to a smaller receiving area, thus resulting in an increased radiation flux. A classification of solar collectors available on the market is provided in Table 1.2.

1.2.1 Concentrating Collectors

When higher temperatures are required, solar radiation needs to be focused. This is achieved by using concentrating collectors. Solar radiation can be concentrated by interposing a reflecting arrangement of mirrors or a refracting arrangement of Fresnel lenses between the source of radiation and the absorber surface.

The optical system which directs the solar radiation on to the absorber is defined concentrator, while the system including the absorber, its cover and other accessories is defined receiver. The reflecting surfaces may be parabolic, spherical or flat and continuous or segmented. Also, they can be classified according to the formation of the image, being either imaging or non-imaging. Imaging concentrators may focus on a line or at a point. The absorber can be convex, flat or concave and it has a reduced area with respect to that of the reflecting/refracting system. In addition, it can be uncovered or surrounded by a transparent cover.

The presence of the optical system compromises the overall performance by adding several inefficiencies:

- reflection losses;
- absorption losses;
- losses due to geometrical imperfections of the optical system.

Table 1.2 Solar thermal energy collectors available in the market

Collector	Tracking	Absorber	T_{max} (°C)
Flat plate	Stationary	Flat	80
Evacuated tube	Stationary	Flat	200
Compound parabolic	Stationary/one-axis	Flat/tubular	250
Parabolic trough	One-axis	Tubular	450
Linear fresnel	One-axis	Tubular	250
Parabolic dish	Two-axis	Point	1500
Heliostat field	Two-axis	Point	2000

However, these inefficiencies are generally compensated for considering that the area from which heat losses occur (i.e., the absorber area) is reduced.

From a technical point of view, concentrating collectors are more complex with respect of stationary collectors since they have to be oriented to track the Sun in order to have the beam radiation directed on to the absorber surface. Different tracking methods are possible and a proper choice depends on the precision with which it has to be done:

- in collectors with a low concentration ratio (see Sect. 2.2.1), it is often adequate to manually make one or two adjustments of the collector orientation every day;
- in collectors with a high concentration ratio, a continuous adjustment of the collector orientation is necessary.

Also, tracking may be required about one or two axis. Obviously, the necessity of a tracking method introduces complexity in the design, and maintenance requirements are also increased; all these factors weight on the costs. In addition, the almost entirely part of diffuse radiation is lost being not focused.

1.2.2 Parabolic Trough Collectors

Parabolic trough collectors (PTCs) are solar devices able to track the sun and to concentrate the solar radiation into a focal line. These collectors can be divided into two distinct parts (Fig. 1.1):

- a concentrator, including the reflector and the support structure;
- a receiver, which includes the absorber tube located at the focal axis through which the HTF flows, and the transparent cover.

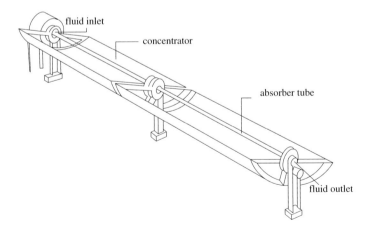

Fig. 1.1 Schematic of a parabolic trough collector (PTC)

The concentrator is a cylindrical parabola which generally presents a reflective foil with high specular reflectance to the solar spectrum. The reflective surface is generally made of a curved back silvered glass. The support structure should not distort significantly due to its own weight and it should be able to withstand wind loads and aggressive atmospheric conditions. The tracking of the sun requires the PTC to rotate about its axis and usually occurs by means of a mechanical transmission system coupled with an electric motor.

The second element of a PTC, the receiver, consists of a metallic absorber tube, generally made of stainless steel, copper or aluminum. It is located at the focal axis of the concentrator and it is coated with a heat resistant black paint to maximize the absorption of the solar radiation. The receiver usually includes a tubular cover transparent to the solar spectrum, useful to reduce convective and radiative heat losses from the absorber to the environment. To improve the performance, the absorber tube can be coated with a selective paint and the annulus between the cover and the absorber can be evacuated.

PTCs are available over a wide range of aperture areas from about 1 to $60\,m^2$, with widths from 1 to $6\,m$ [5]. Fluid temperatures between 50 and $400\,°C$ can be easily reached with this kind of collectors.

1.3 Solar Thermal Applications

Solar energy is used in a great number of thermal applications to meet various energetic needs. These are:

1. water heating;
2. space heating;
3. space cooling and refrigeration;
4. industrial process heat;
5. electric power generation;
6. distillation;
7. drying;
8. cooking.

Although PTCs are usually adopted for medium/high-temperature applications such as industrial processes and electric power generation, actually low-temperature (or low-enthalpy) PTCs are suitable for most of the above applications. The following sections provide an overview of these applications and illustrate how low-temperature PTCs can supply them whenever possible.

1.3.1 Water Heating

Solar water heating is one of the most widespread solar thermal applications and one of the most attractive from an economic standpoint. In many countries of the world, this technology is already competing on equal terms with systems using other energy sources. Hot water is generally used for domestic, industrial and commercial purposes.

Solar water-heating systems are classified into two categories:

- natural circulation (or passive, or thermosyphon) systems;
- forced circulation (or active) systems.

The two main components of a natural circulation system are a liquid solar collector and a storage tank. In addition to these ones, forced circulation systems include a water pump. Due to the need of low temperatures and given their low cost, hot water for domestic purposes is typically provided by flat plate and evacuated tube collectors. These systems can adopt both natural and forced circulation.

On the other hand, when larger amounts of hot water are required to supply industrial or commercial heat demands, a forced circulation system pump is normally adopted and low-enthalpy PTCs could be economically attractive. Solar water systems of this type could be well suited for factories, hospitals, hotel, offices, etc.

1.3.2 Space Heating

Colder countries have a significant demand of thermal energy for space heating. Thus, heat for comfort in buildings can be provided by systems which are similar to water heating systems. These systems can be distinguished into two types:

- active methods;
- passive methods.

Active methods use pumps or blowers to circulate the HTF in the space-heating system. On the other hand, with passive methods the thermal energy flows through a living space by natural means without the help of a mechanical device. Given these applications, solar air heaters are usually adopted for the purpose, since they eliminate the need to transfer heat from one fluid to another.

1.3.3 Space Cooling and Refrigeration

Space cooling can be adopted to provide comfortable living conditions (i.e., air-conditioning) or food preservation (i.e., refrigeration). Since solar energy is received as heat, a smart choice is a system working on an absorption refrigeration cycle which requires most of its energy input as heat. Besides, considering that the generators of commercially available absorption refrigerators require temperatures above 70 °C, PTCs appear very promising for this kind of solar thermal applications.

Unfortunately, the installation cost of a solar absorption refrigeration system is high because of the cost of both the large collector required and the absorption unit itself. Therefore, commercialization has not taken place although a few demonstration units have been set up [5].

1.3.4 Industrial Process Heat

The most important application for solar energy at medium-high temperature (80 ÷ 240 °C) is heat production for industrial processes, which represents a significant amount of heat. Industrial heat demand constitutes about 15% of the overall demand of final energy requirements in the southern European countries [3]. The present heat demand in the European Union for medium and medium-high temperatures is estimated to be about 300 TWh/year [3].

Several industrial sectors have been identified as having favorable conditions for the application of solar energy [3, 6]:

- sterilizing;
- pasteurizing;
- drying of lumber or food;
- hydrolyzing;
- washing;
- cleaning in food processing;
- extraction operations in metallurgical or chemical processing;
- curing of masonry products;
- paint drying;
- polymerization.

Temperatures for these applications can range from near ambient to those corresponding to low-pressure steam. Energy can be provided both from flat plate collectors (for low-temperature applications) and concentrating collectors (for medium-high temperatures). Industries which use most of the energy are the food industry and the manufacture of non-metallic mineral products. Favorable conditions exist in the food industry because food treatment and storage are processes with high energy consumption and running time [3].

Most factories use hot water or steam at a pressure corresponding to the highest temperature required in the different processes. Hot water or low-pressure steam at temperatures lower than 150 °C can be used either for pre-heating water or for steam generation, or by direct coupling of the solar system to an individual process working at temperatures lower than those of the central system supply. One way to ensure low investment costs is to design systems with no heat storage so that the solar heat is used directly into a suitable industrial process. However, this system cannot be cost-effective when heat is required at the early or late hours of the day, or at nighttime [3].

Low-temperature steam/water can be used in industrial applications, sterilization and for powering desalination evaporators. PTCs are frequently adopted for the purpose since suitable temperatures can be obtained with good efficiencies.

1.3.5   Electric Power Generation

The generation of electrical power is one of the most important applications of PTCs. Solar power cycles used to convert solar energy into electric energy can be classified as follows:

- low-temperature cycles ($<100\,°C$);
- medium-temperature cycles ($<400\,°C$);
- high-temperature cycles ($>400\,°C$).

Low-temperature systems generally adopt flat plate collectors working on a Rankine cycle. The overall efficiency of these systems are rather low because the temperature difference between the steam leaving the generator and the condensed liquid leaving the condenser is small. To reduce costs, solar ponds have been proposed. However, these systems are only less costly than plants using flat plate collectors [5].

Solar thermal power plants operating with medium-temperature cycles use PTC technology at temperatures of about $400\,°C$. They have proved to be the most cost effective and successful so far [5]. The first commercial plant of this type was SEGS I (Solar Electric Generating System), a power plant of 14 MWe. It was inaugurated in 1984 by LUZ International Limited in Daggett, in the Mojave Desert in southern California. Since then, six plants of 30 MWe capacity (SEGS II to VII), followed by two plants of 80 MWe (SEGS VIII and IX), were commissioned and installed, for a total installed capacity of 354 MWe. Figure 1.2 shows the aerial view of portions of SEGS III-VII plants.

High-temperature power systems work with paraboloid dish reflectors and heliostat field collectors. Because of the limitations on the size of the concentrator, paraboloid dish reflectors can generate moderate power (of the order of kilowatt). Therefore, they can be expected to meet the power needs of communities, particularly in rural areas [5].

In heliostat field collectors, a HTF flowing through the receiver absorbs the incident radiation and transports the heat to the ground where it is used in a Rankine or a Brayton cycle. Molten salts, water and air have been used as HTFs. Of all the plants built in eighties, the largest (10 MWe) was Solar One, built in 1982 in Barstow, California. The plant has been operated since 1982 for six years. To overcome some problems encountered with Solar One, several modifications were made. Molten salts were used as HTF instead of water/steam so that only a single phase liquid flow occurred in the tubes of the receiver. Also, a new molten salt thermal storage system

Fig. 1.2 Aerial view showing portions of SEGS III-VII plants located at Kramer Junction, California. Photograph by Alan Radecki Akradecki

with a larger capacity was installed and 108 heliostats were added. This modified plant was called Solar Two. The project has been operative since 1996 for three years.

In the last thirteen years, the only plant set up was PS10 (Planta Solar) in Spain. The plant construction was completed in 2006 and the commercial operations started in March 2007. The plant has a power output of 10 MWe.

1.3.6 Distillation

The natural supply of fresh water is inadequate with respect to the availability of brackish water and seawater in many small communities of the world. Therefore, solar distillation could be an effective way of supplying drinking water to such communities and it is no coincidence that it was one of the first attempts made by humanity to exploit solar energy.

Solar distillation systems can be classified into two categories: direct and indirect. The formers collect solar energy to produce distilled water directly in the solar collector, while the latters include two systems: one for the collection of solar energy and one for the distillation process. Among indirect solar distillation systems, low-enthalpy PTCs are generally employed to provide low-pressure steam.

1.3.7 Drying

Drying in agriculture is another one of the traditional uses of solar energy. The aim of a drying process is to reduce the water/moisture contained in agricultural products to prevent a quick deterioration and to guarantee their preservation. Along with distillation, it is one of the oldest applications of solar energy.

Traditionally, drying was done on open ground. The disadvantages associated with this are that the process is slow and that insects and dust get mixed with the product. The use of solar dryers, air collectors which collect solar energy, helps to eliminate these disadvantages: drying can be done faster and a better quality for the product is obtained.

1.3.8 Cooking

The energy demand for cooking in developing countries is an important portion of the global energy consumption. For example, in India about the 50% of thermal energy is used only for cooking, and a large fraction of this demand is satisfied by non-renewable sources such as wood, kerosene and liquefied petroleum gas [7]. Therefore, many countries use particular collectors named solar cookers (or solar ovens) with the aim of cooking different types of food.

Over the past 40 years, a number of designs of solar cookers have been developed. These can be classified into two categories:

- direct, in which solar radiation is directly used to cook food;
- indirect, where solar radiation transfers heat to a HTF which is then delivered to the hob.

Direct solar cookers include box, panel and parabolic type. The formers are easy to realize since they can be manufactured through rudimentary materials such as cardboards, aluminum foils and glue. Solar panel cookers are box cookers which can be integrated with mirrors able to increase the solar energy reaching the absorber. These devices are generally slow (time taken for cooking varies from half an hour to three hours) and suitable for domestic purposes, with temperatures reaching $100\,^\circ$C on clear-sky days.

Solar parabolic cookers are collectors in which the radiation is concentrated by a paraboloid reflecting surface. The cooking vessel is placed at the focus of the paraboloid mirror and temperatures above $200\,^\circ$C can be easily obtained. Thus, they can be used for roasting, frying or boiling food. Unlike box and panel cookers, solar parabolic cookers require a frequent manual tracking in order to work with high efficiency.

As concerns indirect solar cookers, flat plate and evacuated tube collectors are generally used for the collection of solar energy. Evidently, if higher temperatures are required, low-enthalpy PTCs could be adopted for the purpose.

One of the most important limits of solar cookers lies in the fact that they are not normally able to store heat which could be used during nighttime or when it is cloudy. To overcome this limit, several studies have been conduced using materials able to store heat for a long time period. Among these, phase change materials (PCMs) appear to be the most promising. PCMs are substances able to store a large amount of latent heat. Since to date a number of PCMs is commercially available in a very large range of melting temperatures, various types of PCM-based solar cookers could be realized to cook food even when solar radiation is not available.

References

1. Turkenburg WC et al (2000) Renewable energy technologies. Energy and the challenge of sustainability, World energy assessment, pp 219–272
2. Task (2006) Key issues for renewable heat in Europe—solar industrial process heat—state of the art—WP3. 3:5
3. Kalogirou SA (2014) Solar energy engineering: processes and systems, 2nd edn. Elsevier, Oxford
4. Goswami DY, Kreider JF (2000) Principles of solar engineering, 2nd edn. Taylor & Francis, Philadelphia
5. Sukhatme SP, Nayak JK (2008) Solar energy: principles of thermal collection and storage. Tata McGraw-Hill Publishing Company, New Delhi
6. Duffie JA, Beckman WA (2013) Solar engineering of thermal processes, 4th edn. Wiley, Hoboken
7. Cuce E, Cuce PM (2013) A comprehensive review on solar cookers. Appl Ener 102:1399–1421

Chapter 2
Mathematical Modeling

Abstract This chapter presents a detailed mathematical analysis of PTCs. It is divided into three sections: tracking of the Sun, optical analysis and thermal analysis. The first section is an overview of equations and relationships used in solar geometry to determine the position of the Sun and, hence, the slope and the angle of incidence that in each instant must be assumed by a PTC to correctly follow the Sun. Since PTCs usually have one degree of freedom, correlations valid for east-west and north-south axis with continuous adjustment are discussed. The optical analysis starts by introducing the concentration ratio and continues presenting a thorough description of the geometry of a PTC. Optical errors and geometrical effects are also presented. Then, the optical analysis is concluded by taking into account the optical properties of the materials generally adopted in PTCs: the mirror, the cover and the absorber. The last section involves the thermal analysis of a PTC, i.e. the energy balance of the receiver. Each heat flux is described in detail, in order to determine the thermal efficiency of a PTC.

Keywords Tracking · Solar angles · Optical errors · Concentration ratio · Energy balance

2.1 Tracking of the Sun

This section intends to supply the reader with the essential equations to calculate the position of the Sun. Specific equations to be used with PTCs will be also given.

2.1.1 Solar Time

With the term *solar time*, we intend the time based on the apparent angular motion of the Sun across the sky. The time the Sun crosses the meridian of the observer is called *solar noon*. Solar time does not coincide with local clock time (LCT); it is necessary to convert standard time to solar time by applying two corrections [1]:

© The Author(s) 2016
G. Coccia et al., *Parabolic Trough Collector Prototypes for Low-Temperature Process Heat*, SpringerBriefs in Applied Sciences and Technology,
DOI 10.1007/978-3-319-27084-5_2

1. the first correction is for the difference in longitude between the observer's meridian and the meridian on which the local standard time is based;
2. the second correction derives from the equation of time, which accounts for the perturbations in the Earth's rate of rotation that affects the time the Sun crosses the observer's meridian.

The difference expressed in minutes between solar time and standard time is [1]:

$$\text{Solar Time} - \text{Standard Time} = 4\,(L_{st} - L_{loc}) + E \tag{2.1}$$

where L_{st} is the standard meridian for the local time zone and L_{loc} is the longitude of the site (longitudes are in degrees west). The symbol E denotes the equation of time, expressed in minutes [1]:

$$E = 229.2\,(0.000075 + 0.001868 \cos B - 0.032077 \sin B$$
$$- 0.014615 \cos 2B - 0.04089 \sin 2B) \tag{2.2}$$

where

$$B = (n - 1)\,\frac{360}{365} \tag{2.3}$$

In Eq. (2.3), n is the nth day of the year. Note that Eq. (2.2) and all the equations in the following use degrees and not radians. Time is assumed to be solar time unless indication is given otherwise.

2.1.2 Solar Angles

The Sun's position with respect to a plane of any particular orientation to the Earth at any time can be described in terms of several angles, reported in Fig. 2.1. The angles and their sign conventions are as follows [1].

- Latitude (ϕ): it is the angular location of a terrestrial site with reference to the Equator. It is north positive: $-90° < \phi < 90°$.
- Declination (δ): it is the angular position of the Sun at solar noon with respect to the plane of the Equator. It is north positive: $-23.45° < \delta < 23.45°$. The declination can be found with an error less than $0.035°$ with the following equation [1]:

$$\delta = (180/\pi)(0.006918 - 0.399912 \cos B + 0.070257 \sin B$$
$$- 0.006758 \cos 2B + 0.000907 \sin 2B$$
$$- 0.002697 \cos 3B + 0.00148 \sin 3B) \tag{2.4}$$

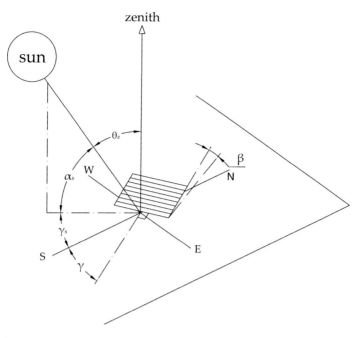

Fig. 2.1 Solar angles for a tilted surface. Adapted from [1]

where B was defined in Eq. (2.3).

- Slope (β): it is the angle between the plane of a surface and the horizontal. If it is greater than 90°, the surface has a downward-facing component: $0° < \beta < 180°$.
- Surface azimuth angle (γ): it is the deviation of the projection on a horizontal plane of the normal to the surface from the local meridian. It is zero due south, east negative and west positive: $-180° < \gamma < 180°$.
- Hour angle (ω): it is the angular displacement of the Sun east or west of the local meridian due to rotation of the Earth on its axis at 15° per hour. It is negative in the morning and positive in the afternoon.
- Angle of incidence (θ): it is the angle between the beam radiation on a surface and the normal to that surface. The angle of incidence is related to the above mentioned angles through the expression:

$$\cos \theta = \sin \delta \sin \phi \cos \beta - \sin \delta \cos \phi \sin \beta \cos \gamma$$
$$+ \cos \delta \cos \phi \cos \beta \cos \omega + \cos \delta \sin \beta \sin \gamma \sin \omega$$
$$+ \cos \delta \sin \phi \sin \beta \cos \gamma \cos \omega \qquad (2.5)$$

The angle of incidence may exceed 90°, meaning that the Sun is behind the surface. Equation (2.5) implies that the hour angle is between sunrise and sunset.

- Zenith angle (θ_z): it is the angle between the vertical and the line of the Sun. If the surface is horizontal ($\beta = 0$), θ_z corresponds to the angle of incidence and

Eq. (2.5) becomes:

$$\cos \theta_z = \sin \delta \sin \phi + \cos \delta \cos \phi \cos \omega \tag{2.6}$$

- Solar altitude angle (α_s): it is the angle between the horizontal plane and the beam radiation. It is the complementary angle of the zenith.
- Solar azimuth angle (γ_s): it is the angular displacement from south of the projection of the line of the Sun on the horizontal plane. Displacements east of south are negative and west of the south are positive. The solar azimuth angle can be found with:

$$\gamma_s = \text{sgn}(\omega) \left| \cos^{-1} \left(\frac{\cos \theta_z \sin \phi - \sin \delta}{\sin \theta_z \cos \phi} \right) \right| \tag{2.7}$$

where the sign function is equal to $+1$ if ω is positive and to -1 if ω is negative.

2.1.3 Angles for Tracking Surfaces

Solar collectors such as PTCs track the Sun by moving in prescribed ways to minimize the angle of incidence of beam radiation on their surface and therefore maximize the incident direct radiation. In particular, PTCs can rotate about their axis that could have any orientation but in practice is usually horizontal east-west or horizontal north-south.

For a plane rotated about a horizontal east-west axis with continuous adjustment to minimize the angle of incidence [1]:

$$\cos \theta = \sqrt{1 - \cos^2 \delta \sin^2 \omega} \tag{2.8}$$

The slope of the surface can be calculated from:

$$\tan \beta = \tan \theta_z \ |\cos \gamma_s| \tag{2.9}$$

If the solar azimuth angle passes through $\pm 90°$, the surface azimuth angle of orientation will change between $0°$ and $180°$; otherwise:

$$\gamma = \begin{cases} 0°, & \text{if } |\gamma_s| < 90° \\ 180°, & \text{if } |\gamma_s| \geq 90° \end{cases} \tag{2.10}$$

The shadowing effects of this arrangement are minimal; the principal shadowing is caused when the collector is tipped to a maximum degree south ($\delta = 23.5°$) at winter solstice. In this case, the Sun casts a shadow toward the collector at the north. This configuration has the advantage to approximate the full tracking in summer;

however, the winter performance is seriously depressed relative to the summer one [2].

For a plane rotated about a horizontal north-south axis with continuous adjustment to minimize the angle of incidence [1]:

$$\cos \theta = \sqrt{\cos^2 \theta_z + \cos^2 \delta \sin^2 \omega} \tag{2.11}$$

The slope is:

$$\tan \beta = \tan \theta_z \; |\cos (\gamma - \gamma_s)| \tag{2.12}$$

In this arrangement, γ will be 90° or −90° depending on the sign of γ_s:

$$\gamma = \begin{cases} 90°, & \text{if } \gamma_s > 0° \\ -90°, & \text{if } \gamma_s \leq 0° \end{cases} \tag{2.13}$$

The greatest advantage of this arrangement is that very small shadowing effects are encountered when more than one collector is used. These occur only at the first and last hours of the day [2].

2.1.4 Beam Radiation on Tilted Surfaces

According to the instrument used to measure solar radiation, different relationships should be considered to identify the beam radiation which falls on a tilted surface.

Pyrheliometers are instruments able to measure the normal beam radiation, G_{bn}. Therefore, if G_{bn} measurements are available, the beam radiation on a tilted surface is (see Fig. 2.2):

$$G_{bt} = G_{bn} \cos \theta \tag{2.14}$$

On the other hand, pyranometers measure global (i.e., direct and diffuse) solar radiation referred to the horizontal plane, G. The same quantity is generally considered in estimates of solar radiation given by empirical equations. If we only consider the direct fraction G_b, from Fig. 2.2 we obtain that:

$$G_{bt} = G_{bn} \cos \theta = G_b \frac{\cos \theta}{\cos \theta_z} \tag{2.15}$$

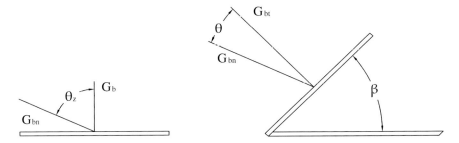

Fig. 2.2 Beam radiation on horizontal and tilted surfaces. Adapted from [1]

2.2 Optical Analysis

The optical analysis quantifies the amount of solar energy that actually reaches a PTC receiver. In the following sections, the parameters that influence a PTC optical efficiency will be analyzed.

2.2.1 Concentration Ratio

In PTCs, the concentration of solar radiation is achieved by reflecting the solar flux incident on the concentrator of aperture area A_a onto the receiver of area A_r. The concentration ratio, C, is referred to as the ratio of the aperture area to that of the receiver:

$$C = \frac{A_a}{A_r} \tag{2.16}$$

Generally, the higher the temperature at which energy is to be delivered, the higher should be the concentration ratio. This ratio has also an upper limit that depends on the second law of thermodynamics. Since PTCs are two-dimensional concentrating collectors, it is possible to demonstrate that the maximum achievable concentration ratio is [3]:

$$C_{max} = \frac{1}{\sin \theta_m} \tag{2.17}$$

where θ_m is the acceptance half-angle the Sun subtends as seen from the Earth. For tracking collectors, θ_m is limited by the size of the Sun's disk, small-scale errors, irregularities of the reflector surface and tracking errors [2]. For a perfect PTC, C_{max} depends only on the Sun's disk. In this case, the half-acceptance angle is $0.267°$ and we get:

$$C_{max} \simeq 215 \tag{2.18}$$

2.2.2 Geometry of a PTC

The cross-section of a PTC is shown in Fig. 2.3. If x is the horizontal axis and y is the vertical axis, the equation of the parabola is:

$$y = \frac{1}{4f}x^2 \tag{2.19}$$

where f is the focal length of the parabola, the distance from the focal point to the vertex.

The radiation beam of Fig. 2.3 is incident at the rim of the concentrator. The angle ϕ_r, made by the reflected beam radiation with the center line, is called rim angle:

$$\phi_r = \tan^{-1}\left[\frac{8(f/W_a)}{16(f/W_a)^2 - 1}\right] = \sin^{-1}\left(\frac{W_a}{2r_r}\right) \tag{2.20}$$

where W_a is the aperture of the parabola and r_r is the maximum mirror radius equal to

$$r_r = \frac{2f}{1 + \cos\phi_r} \tag{2.21}$$

Equation (2.20) can be rearranged to find an expression for the aperture:

$$W_a = 2r_r \sin\phi_r = \frac{4f \sin\phi_r}{1 + \cos\phi_r} = 4f \tan\frac{\phi_r}{2} \tag{2.22}$$

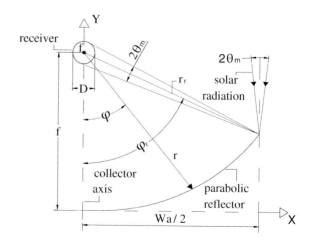

Fig. 2.3 Cross-section of a PTC. Adapted from [2]

The equation of the parabola can be integrated from 0 to $W_a/2$ (or from 0 to ϕ_r) to find its arc length:

$$L_p = \frac{f}{2}\left[\tan\frac{\phi_r}{2}\sec\frac{\phi_r}{2} + \ln\left(\tan\frac{\phi_r}{2} + \sec\frac{\phi_r}{2}\right)\right] \qquad (2.23)$$

For specular reflectors of perfect alignment, the minimum size of the receiver of diameter D necessary to intercept all the reflected radiation is:

$$D = 2r_r\sin\theta_m \qquad (2.24)$$

The proper value of the half-acceptance angle θ_m used in Eq. (2.24) depends on the accuracy of the tracking mechanism and the irregularities of the reflector surface. The smaller these two effects, the closer is θ_m to the Sun's disk angle, resulting in a smaller image and higher concentration. In Fig. 2.3, the incident beam of solar radiation is a cone with an angular width of $0.53°$ (a half-angle θ_m of $0.267°$); it leaves the concentrator at the same angle. This situation occurs only with a perfect PTC. With a real PTC, the half-acceptance angle should be increased to include the presence of errors [2]. All these are accounted for by the intercept factor, which will be discussed in Sect. 2.2.3.

For a tubular receiver of the same length of the reflector, the concentration ratio is:

$$C = \frac{W_a}{\pi D} \qquad (2.25)$$

Substituting Eqs. (2.22) and (2.24) into Eq. (2.25):

$$C = \frac{\sin\phi_r}{\pi\sin\theta_m} \qquad (2.26)$$

C is maximum when $\sin\phi_r = 1$ (i.e., when $\phi_r = 90°$). Therefore, Eq. (2.26) becomes:

$$C_{max} = \frac{1}{\pi\sin\theta_m} \qquad (2.27)$$

The difference between this equation and Eq. (2.17) is that the former applies to a PTC with a circular receiver, while the latter refers to the idealized case. In comparison with Eq. (2.18), and using the same half-acceptance angle of $0.267°$, the maximum concentration ratio is $C_{max} = 1/(\pi\sin 0.267°) = 68.3$.

Finally, it can be demonstrated that, with $\phi_r = 90°$, the mean focus-to-reflector distance and the reflected beam are minimized, so that the slope and tracking errors are less pronounced [4]. The collector's surface area, however, decreases as the rim angle decreases. Thus, there is a temptation to use smaller rim angles because the reduction in optical efficiency is small in comparison with the saving in reflective material cost [2].

2.2.3 Optical Errors

The upper limit to the concentration ratio which can be achieved by a PTC is set by the Sun's width, as seen in Sect. 2.2.1. However, in practical use, the concentration ratio is degraded below to this upper limit due to several factors:

- apparent changes in Sun's width and incidence angle effects;
- physical properties of the materials used in the construction;
- imperfections that may result from manufacture and/or assembly, imperfect tracking of the Sun, and poor operating procedures.

A depth study of all potential errors in PTCs was presented by Guven and Bannerot [5]. Errors can be divided into two groups: random errors and non-random errors (Fig. 2.4). Random errors are defined as truly random natural errors and, therefore, can be represented by normal distributions with mean equal to zero. They are treated statistically and are the origin of spreading of the reflected energy distribution. Random errors are:

- scattering effects associated with the optical material used in the reflector;
- scattering effects caused by random slope errors (e.g., waviness of the reflector due to distortions occurred during manufacturing and/or assembly);
- misalignment of the PTC with the Sun due to random tracking errors (which last only a very short period of time).

These errors can be modeled statistically by introducing a total reflected energy distribution standard deviation at normal incidence, $\sigma_{tot,n}$, which is given by:

$$\sigma_{tot,n} = \sqrt{\sigma_{sun,n}^2 + \sigma_{mirror,n}^2 + 4\sigma_{slope,n}^2} \tag{2.28}$$

In this equation:

- $\sigma_{sun,n}$ is the energy distribution standard deviation of the Sun's rays at normal incidence and solar noon;
- $\sigma_{mirror,n}$ is the standard deviation of the distribution of diffusivity of the reflective material at normal incidence;
- $\sigma_{slope,n}$ is the standard deviation of the distribution of local slope errors at normal incidence.

Non-random errors have a single deterministic value and can be related directly to anticipated errors in manufacture/assembly and/or in operation. In general, these errors will cause the central ray of the reflected energy distribution to shift from the design direction. Non-random errors can be classified as:

- Reflector profile errors (e.g., due to deflection or severe waviness of the reflector surface) which cause a permanent change in the location of the focus of the reflector, thus preventing the reflected radiation to reach the receiver. They can be quantified with the distance between the actual and ideal focus measured along the optical axis of the reflector.

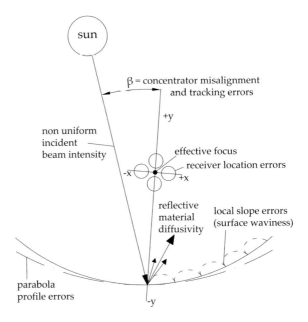

Fig. 2.4 Optical errors in a PTC. Adapted from [5]

- Misalignment of the trough with the Sun (e.g., due to a constant tracking error) so that the position of the focus is shifted from the ideal focus and the central ray of the reflected beam can miss the receiver. The distance between the ideal focus of the concentrator and the center of the receiver can be used to quantify them.
- Misalignment of the receiver with the effective focus of the concentrator, that causes the central ray to miss the receiver. This quantity can be evaluated by defining an angle between the central solar ray and the normal to the concentrator aperture plane, β, as shown in Fig. 2.4.

Guven and Bannerot [5] showed that the receiver mislocation along the optical axis (y axis) degrades the optical performances more than the mislocation along the x axis (Fig. 2.4). Thus, the receiver mislocation along the optical axis, $(d_r)_y$, can be chosen to represent the non-random receiver location errors. Since the reflector profile errors and the receiver mislocation along y axis bring about the same effect, the parameter $(d_r)_y$ can account for both. Therefore, only two independent variables, $(d_r)_y$ and β, are sufficient to model non-random errors.

In summary, there are three error parameters that characterize optical errors: one random error, described by $\sigma_{tot,n}$, and two non-random errors, described by $(d_r)_y$ and β, respectively. To quantify all the errors with a single parameter, the intercept factor is introduced. This is referred to as the fraction of reflected radiation that is incident on the absorbing surface of the receiver and it is a function of both random and non-random errors as well as the geometry of the collector:

$$\gamma = \gamma \left(\phi_r, C, D, \sigma_{tot,n}, (d_r)_y, \beta \right) \tag{2.29}$$

Random and non-random errors can be combined with the geometrical parameters of the PTC to conduct an analysis valid for all PTC geometries [5]. The expression of γ derived by Guven and Bannerot [5] is:

$$\gamma = \frac{1 + \cos \phi_r}{2 \sin \phi_r}$$
$$\times \int_0^{\phi_r} \left\{ erf\left(\frac{\sin \phi_r(1 + \cos \phi)(1 - 2d^* \sin \phi) - \pi \beta^*(1 + \cos \phi_r)}{\sqrt{2} \pi \sigma^*(1 + \cos \phi_r)} \right) \right.$$
$$\left. - erf\left(-\frac{\sin \phi_r(1 + \cos \phi)(1 + 2d^* \sin \phi) + \pi \beta^*(1 + \cos \phi_r)}{\sqrt{2} \pi \sigma^*(1 + \cos \phi_r)} \right) \right\}$$
$$\times \frac{d\phi}{1 + \cos \phi} \tag{2.30}$$

where

- $\sigma^* = \sigma_{tot,n} C$ is the universal random error parameter;
- $d^* = (d_r)_y/D$ is the universal non-random error parameter due to receiver mislocation and reflector profile errors;
- $\beta^* = \beta C$ is the universal non-random error due to angular errors.

Equation (2.29) can be therefore simplified as:

$$\gamma = \gamma \left(\phi_r, \sigma^*, d^*, \beta^* \right) \tag{2.31}$$

2.2.4 Geometrical Effects

Several abnormal incidence factors have the effect of reducing the optical performances of a PTC. These factors are:

- end effects;
- shading by integral bulkheads;
- intra-array shading.

Jeter et al. [6] presented a technique which ascribes to these effects the purely geometrical result of reducing the effective aperture of the concentrator.

During off-normal operation of a PTC, some of the rays reflected from near the end of the concentrator cannot reach the receiver. This loss of effective aperture is called end effect. PTCs can exhibit end effects since the receiver is usually terminated near the same cross-section plane as in the concentrator. The effective area lost to end effects is represented by the ruled region in Fig. 2.5 and is equal to [6]:

$$A_i = fw \tan \theta \left(1 + \frac{w^2}{48f^2} \right) \tag{2.32}$$

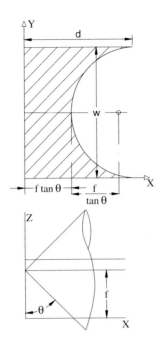

Fig. 2.5 Ineffective aperture area due to end effects. Adapted from [6]

where f is the focal distance, w is the parabola width and θ is the angle of incidence.

It is possible to reduce end effects by employing a receiver longer than the trough. If this solution is adopted, two cases must be considered, as Fig. 2.6 shows. If $S < A$ (Fig. 2.6a), the ineffective area is:

$$A_{i,1} = A_i - Sw \tag{2.33}$$

where S is the distance between the receiver rim and the concentrator rim. Instead, if $A < S < d$ (Fig. 2.6b), it can be demonstrated that the ineffective area is given by:

$$A_{i,2} = A_i + (Sw' - A_i') - Sw \tag{2.34}$$

Clearly, if $S > d$, the ineffective area is equal to zero.

Combining the end effects with the shading produced by the receiver, the ratio of ineffective area to the whole aperture area is:

$$A_f = \frac{A_i + D_{ao}L_c}{A_a} \tag{2.35}$$

where D_{ao} is the cover outer diameter and L_c is the length of the concentrator. Thus, the effective aperture area is:

$$A_{ae} = A_a(1 - A_f) \tag{2.36}$$

Fig. 2.6 End effects when
the receiver extends beyond
the trough. **a** $S < A$.
b $A < S < d$

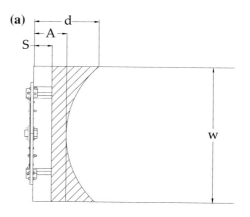

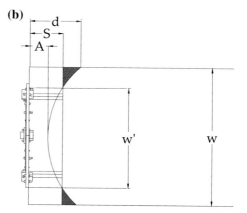

2.2.5 Optical Properties of Materials

The following sections present equations to calculate the optical properties of
the materials adopted in a PTC, in particular the absorptance of the absorber and
the transmittance of the glass cover.

2.2.5.1 Specular Reflectance of the Mirror

PTCs require the use of reflecting materials such to direct the beam radiation onto
the receiver. Therefore, surfaces of high specular reflectance for radiation in the solar
spectrum are required. Specular surfaces are usually metals or metallic coatings on
smooth substrates. The specular reflectivity of such surfaces is a function of the
quality of the substrate and the plating.

Specular reflectance generally depends on wavelength, so monochromatic reflectances should be integrated for the particular spectral distribution of incident beam radiation. Typical values of specular reflectance of surfaces for solar radiation are greater than 0.90.

2.2.5.2 Glass Cover

The transmittance of the glass cover of a PTC can be obtained with adequate accuracy by considering reflection and absorption separately, and is given by the product form:

$$\tau \simeq \tau_r \tau_a \tag{2.37}$$

where τ_r is the transmittance obtained by considering only reflection losses and τ_a is the transmittance obtained by considering only absorption losses.

The transmittance τ_r can be evaluated from:

$$\tau_r = \frac{1}{2}\left(\frac{1-r_\perp}{1+r_\perp} + \frac{1-r_\parallel}{1+r_\parallel}\right) \tag{2.38}$$

where $r_\perp$ and $r_\parallel$ are, respectively, the perpendicular and parallel components of the unpolarized radiation. Those components are given by the Fresnel's equations:

$$r_\perp = \frac{\sin^2(\theta_2 - \theta_1)}{\sin^2(\theta_2 + \theta_1)} \tag{2.39}$$

$$r_\parallel = \frac{\tan^2(\theta_2 - \theta_1)}{\tan^2(\theta_2 + \theta_1)} \tag{2.40}$$

In Fresnel's equations, θ_1 is the angle of incidence and θ_2 is the angle of refraction, as depicted in Fig. 2.7. The two angles are related by the Snell's law:

$$\frac{\sin\theta_1}{\sin\theta_2} = \frac{n_2}{n_1} \tag{2.41}$$

where n_1 and n_2 are the refraction indices of the two media forming the interface. Typical values of the refraction index are 1 for air and 1.526 for glass.

The absorption of radiation in a partially transparent medium is described by the Bouguer's law:

$$\tau_a = \exp\left(-\frac{Kt}{\cos\theta_2}\right) \tag{2.42}$$

where K is the extinction coefficient, which is assumed to be a constant in the solar spectrum, and t is the thickness of the glass cover. For glass, the value of K can vary from $4\,\mathrm{m}^{-1}$ (high-quality glass) to approximately $32\,\mathrm{m}^{-1}$ (low-quality glass).

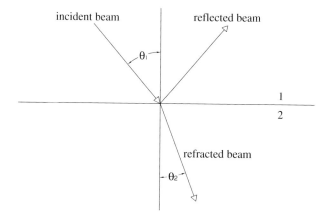

Fig. 2.7 Reflection and refraction at the interface of two media

The absorptance of the cover can be calculated by the following approximate equation:

$$\alpha_c \simeq 1 - \tau_a \tag{2.43}$$

The reflectance of the cover can be found from:

$$\rho_c = 1 - (\alpha_c + \tau) \simeq \tau_a - \tau_a \tau_r = \tau_a(1 - \tau_r) \tag{2.44}$$

2.2.5.3 Absorptance of the Absorber

The absorptance for solar radiation of ordinary blackened surfaces is a function of the angle of incidence of the radiation on the surface. However, the angular dependence of solar absorptance of most surfaces used for solar collectors is not available. An example of this dependence, valid for 0°–90°, is [1]:

$$\begin{aligned}
\alpha/\alpha_n = {} & 1 - 1.5879 \times 10^3 \theta + 2.7314 \times 10^{-4} \theta^2 \\
& - 2.3026 \times 10^{-5} \theta^3 + 9.0244 \times 10^{-7} \theta^4 \\
& - 1.8000 \times 10^{-8} \theta^5 + 1.7734 \times 10^{-10} \theta^6 \\
& - 6.9937 \times 10^{-13} \theta^7
\end{aligned} \tag{2.45}$$

where α_n is the solar absorptance at normal incidence and θ, the angle of incidence, is in degrees.

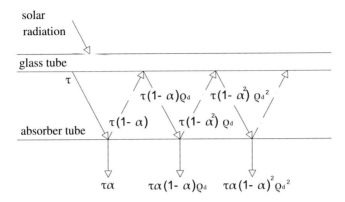

Fig. 2.8 Radiation transfer between the glass and the absorber

2.2.5.4   Transmittance-Absorptance Product

Part of the radiation passing through the cover and incident on the absorber is reflected back to the cover. However, all this radiation is not lost because a portion of it is reflected back to the absorber.

The situation is shown in Fig. 2.8. The fraction $\tau\alpha$ of the incident beam radiation is absorbed by the absorber and the fraction $(1-\alpha)\tau$ is reflected back to the cover. This radiation, that is assumed to be diffuse and unpolarized, reaches the cover and a fraction $(1-\alpha)\tau\rho_{\mathrm{d}}$ is reflected back to the absorber. The term ρ_{d} represents the reflectance of the cover system for diffuse radiation incident from the bottom side and can be evaluated from Eq. (2.44) at an angle of 60°.[1] The multiple reflection of diffuse radiation continues so that the fraction of the incident energy absorbed is

$$(\tau\alpha) = \tau\alpha \sum_{n=0}^{\infty} \left[(1-\alpha)\rho_{\mathrm{d}} \right]^{n} = \frac{\tau\alpha}{1 - (1-\alpha)\rho_{\mathrm{d}}} \tag{2.46}$$

The term $(\tau\alpha)$ is usually referred to as the transmittance-absorptance product. It is possible to prove that a reasonable approximation of Eq. (2.46) for most practical solar collectors is [1]:

$$(\tau\alpha) \simeq 1.01\tau\alpha \tag{2.47}$$

---

[1]For a wide range of conditions encountered in solar collector applications, the equivalent angle for beam radiation, i.e. the angle which gives the same reflectance as for diffuse radiation, is essentially 60° [1].

2.3 Thermal Analysis

In this section, we offer a detailed overview of the heat transfer mechanisms participating in a PTC receiver. The definition of thermal efficiency is also given.

2.3.1 Energy Balance of the Receiver

The thermal performance of a PTC can be evaluated by an energy balance that determines the fraction of the incoming radiation delivered as useful energy to the heat transfer fluid (HTF). A number of simplifying assumptions are usually adopted to model such systems [7]:

- Thermal performances are evaluated under steady-state conditions.
- Heat transfer is one-dimensional, i.e. it occurs only through the receiver radial direction. Note that the assumption of one-dimensional energy balance gives reasonable results for short receivers (<100 m), but it is inadequate for longer receivers [8].
- The thermophysical and optical properties of materials are independent of temperature.
- Heat losses through support brackets are neglected.
- The sky can be considered as a blackbody at an equivalent sky temperature for long-wavelength radiation.
- The effects of dust and dirt are negligible.

Figure 2.9 shows the one-dimensional steady-state energy balance for the receiver cross-section of a PTC, while Fig. 2.10 shows the thermal resistance model. When the beam radiation reflected by the concentrator (S_c) strikes the cover, a fraction of solar energy is transmitted to the absorber (τS_c). Only a portion of this energy, S, is effectively conducted through the absorber ($Q_{k,a}$) and transferred to the HTF by convection ($Q_{c,af}$), while a significant portion is lost and transmitted back by convection ($Q_{c,ac}$) and radiation ($Q_{r,ac}$). The energy lost by convective and radiative heat transfers is transmitted by conduction through the cover ($Q_{k,c}$) and, along with the energy absorbed by the cover ($\alpha_c S_c$), is lost to the environment by convection ($Q_{c,ce}$) and radiation ($Q_{r,ce}$).

   The system of energy-balance equations is determined by applying the conservation of energy at each surface of the receiver cross-section in Fig. 2.9:

$$\begin{cases} S = Q_{k,a} + Q_{c,ac} + Q_{r,ac} \\ Q_{k,a} = Q_{c,af} = Q_u \\ Q_{c,ac} + Q_{r,ac} = Q_{k,c} \\ Q_{k,c} + \alpha_c S_c = Q_{c,ce} + Q_{r,ce} \end{cases} \qquad (2.48)$$

The description of all terms in System (2.48) is provided in Table 2.1.

Fig. 2.9 Energy balance for
the receiver cross-section of
a PTC. The definition of the
symbols is provided in
Table 2.1

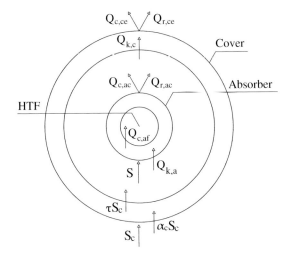

Fig. 2.10 Thermal
resistance model of the
receiver

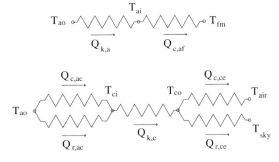

The energy-balance system assumes the contribution of the diffuse component of solar radiation to be negligible.[2] With this assumption, the solar beam radiation reflected by the concentrator to the receiver is:

$$S_c = \rho \gamma G_{bt} A_{ae} \qquad (2.49)$$

where

- ρ is the specular reflectance of the concentrator;
- γ is the intercept factor given by Eq. (2.30);
- G_{bt} is the beam radiation measured on the plane of aperture, it can be evaluated with Eqs. (2.14) and (2.15);
- A_{ae} is the effective aperture area defined in Eq. (2.36).

---

[2]This assumption is acceptable for all concentrators expect for those with low concentration ratio (i.e., for $C = 10$ or below). For systems with low concentration ratio, part of the diffuse radiation will be reflected to the receiver, with the amount depending on the acceptance angle of the concentrator [1].

Table 2.1 Heat fluxes involved in the energy balance of the receiver

Heat flux	Description
S_c	Beam radiation reflected towards the receiver
$\alpha_c S_c$	Beam radiation absorbed by the cover
S	Beam radiation collected by the absorber
$Q_{k,a}$	Conduction through the absorber
$Q_{c,af}$	Convection from the absorber to the fluid
Q_u	Useful heat gain of the fluid
$Q_{c,ac}$	Convection loss from the absorber to the cover
$Q_{r,ac}$	Radiation loss from the absorber to the cover
$Q_{k,c}$	Conduction loss through the cover
$Q_{c,ce}$	Convection loss from the cover to the environment
$Q_{r,ce}$	Radiation loss from the cover to the environment

Thus, the solar beam radiation collected in the absorber is[3]:

$$S = (\tau\alpha)S_c = (\tau\alpha)\rho\gamma G_{bt}A_{ac} \tag{2.50}$$

where the term $(\tau\alpha)$ is the transmittance-absorptance product defined in Eq. (2.46). The following sections describe the involved heat fluxes in detail.

2.3.1.1   Conduction Through the Absorber

The conductive heat transfer through the absorber is given by Fourier's law for concentric cylinders:

$$Q_{k,a} = \frac{2\,\pi\,\lambda_a L(T_{ao} - T_{ai})}{\ln\,(D_{ao}/D_{ai})} \tag{2.51}$$

where

- λ_a is the thermal conductivity of the absorber;
- L is the receiver length;
- T_{ao} is the outer absorber temperature;
- T_{ai} is the inner absorber temperature;
- D_{ao} is the outer absorber diameter;
- D_{ai} is the inner absorber diameter.

Note that λ_a is constant and independent of temperature.

---

[3]One should also consider the solar beam radiation which falls directly on the absorber tube, but this contribution can be ignored when the concentration ratio is high [9].

2.3.1.2 Internal Convection

The heat transfer between the absorber and the HTF occurs by forced convection and can be expressed by the Newton's law:

$$Q_{c,af} = h_f \, \pi \, D_{ai} L (T_{ai} - T_{fm}) \tag{2.52}$$

where T_{fm} is the mean fluid temperature. The convective heat transfer coefficient of the fluid is defined as:

$$h_f = \frac{Nu_f \lambda_f}{D_{ai}} \tag{2.53}$$

where

- Nu_f is the Nusselt number of the HTF;
- λ_f is the thermal conductivity of the HTF.

Note that λ_f is evaluated at the mean fluid temperature.

The Nusselt number depends on the type of flow through the absorber: laminar or transitional/turbulent. For a fluid circulating in a pipe, the flow can be considered laminar when the Reynolds number is lower than 2300. In this condition, the Nusselt number is independent of Reynolds and Prandtl numbers and assumes a constant value equal to 4.36.

On the other hand, the flow of the HTF is within turbulent flow region when $Re_f > 4000$. If $Re_f > 2300$, the Gnielinski's correlation [10] can be used:

$$Nu_f = \frac{(f/8)\,(Re_f - 1000)\,Pr_f}{1 + 12.7(f/8)^{1/2}\left(Pr_f^{2/3} - 1\right)} \tag{2.54}$$

Equation (2.54) is valid for $0.5 \leq Pr_f \leq 2000$ and $2 \times 10^3 < Re_f < 5 \times 10^6$. The Reynolds and Prandtl numbers must be evaluated at the mean fluid temperature. The friction factor f can be estimated from the Colebrook's iterative formula [11]:

$$\frac{1}{\sqrt{f}} = -2 \log \left(\frac{\xi/D_{ai}}{3.71} + \frac{2.51}{Re_f \sqrt{f}} \right) \tag{2.55}$$

where ξ is the pipe roughness.

2.3.1.3 Convective Loss in the Annulus

The heat lost by convection between the absorber and the cover differs if the annulus is either evacuated or not. In the first case, heat transfer occurs by free-molecular convection; in the second case, heat flux is given by free convection. When the receiver annulus is under vacuum (i.e., when the pressure is lower than 1 torr), free-molecular convection can be evaluated as [12]:

$$Q_{c,ac} = h_{ann} \, \pi \, D_{ao}L(T_{ao} - T_{ci})$$ (2.56)

and

$$h_{ann} = \frac{\lambda_{std}}{D_{ao}/2 \ln (D_{ci}/D_{ao}) + bk(D_{ao}/D_{ci} + 1)}$$ (2.57)

$$b = \frac{(2 - a)(9\gamma - 5)}{2a(\gamma + 1)}$$ (2.58)

$$k = \frac{2.331 \times 10^{-20} \left[(T_{ao} + T_{ci})/2 + 273.15\right]}{p\delta^2}$$ (2.59)

where

- h_{ann} is the convective heat transfer coefficient of the annulus gas;
- T_{ci} is the inner cover temperature;
- λ_{std} is the thermal conductivity of the annulus gas at standard temperature and pressure;
- D_{ci} is the inner cover diameter;
- b is the interaction coefficient;
- k is the mean-free path between collisions of a molecule;
- a is the accommodation coefficient;
- γ is the ratio of specific heats for the annulus gas;
- p is the annulus gas pressure;
- δ is the molecular diameter of the annulus gas.

Equation (2.56) can be used when $Ra_{ann} < [D_{ci}/(D_{ci} - D_{ao})]^4$, where Ra_{ann} is the Rayleigh number of the annulus gas. The constants for air as annulus gas are provided in Table 2.2.

If the receiver annulus is not evacuated, heat transfer between the absorber and the cover occurs by free convection:

$$Q_{c,ac} = \frac{2 \, \pi \, \lambda_{eff}L}{\ln (D_{ci}/D_{ao})}(T_{ao} - T_{ci})$$ (2.60)

A recommended correlation for the effective conductive coefficient, λ_{eff}, is [13]:

$$\frac{\lambda_{eff}}{\lambda_{ann}} = 0.386 \left(\frac{Pr_{ann}}{0.861 + Pr_{ann}}\right) \left(F_{cyl}Ra_{ann}\right)$$ (2.61)

Table 2.2 Constants for air as annulus gas at $T_{fm} = 300\,^{\circ}\text{C}$ [14]

$\lambda_{std}\ (\text{Wm}^{-1}\text{K}^{-1})$	k (m)	δ (m)	b	γ
0.02551	0.8867	3.53×10^{-6}	1.571	1.39

where

- λ_{ann} is the thermal conductivity of air evaluated at mean temperature $(T_{ao} + T_{ci})/2$;
- Pr_{ann} is the Prandtl number of air evaluated at mean temperature $(T_{ao} + T_{ci})/2$;
- Ra_{ann} is the Rayleigh number of air evaluated at mean temperature $(T_{ao} + T_{ci})/2$ and characteristic length $(D_{ci} - D_{ao})/2$.

The form factor for concentric cylinders is given by:

$$F_{cyl} = \frac{[\ln (D_{ci}/D_{ao})]^4}{[(D_{ci} - D_{ao})/2]^3 \left(D_{ci}^{-3/5} + D_{ao}^{-3/5}\right)^5} \tag{2.62}$$

Equation (2.61) is valid for $0.70 \leq Pr_{ann} \leq 6000$ and for $10^2 < F_{cyl}Ra_{ann} < 10^7$. Note that if $F_{cyl}Ra_{ann} < 10^2$, convection is negligible and $\lambda_{eff} = \lambda_{ann}$. Finally, λ_{eff} cannot be less than λ_{ann}, so one should set the last equivalence if $\lambda_{eff} < \lambda_{ann}$.

2.3.1.4 Radiative Loss in the Annulus

The heat transfer by radiation between the absorber and the cover can be evaluated with the expression:

$$Q_{r,ac} = \frac{\pi D_{ao}L\,\sigma(T_{ao}^4 - T_{ci}^4)}{1/\varepsilon_a + (1 - \varepsilon_c)(D_{ao}/D_{ci})/\varepsilon_c} \tag{2.63}$$

where

- σ is the Stefan–Boltzmann constant ($5.67 \times 10^{-8}\,\text{Wm}^{-2}\,\text{K}^{-4}$);
- ε_a is the emissivity of the absorber in the long-wavelength range;
- ε_c is the emissivity of the cover in the long-wavelength range.

In Eq. (2.63), temperatures are in kelvin and emissivities assume constant values.

2.3.1.5 Conductive Loss Through the Cover

The heat transfer mechanism described for the absorber is still valid for the cover. Equation (2.51) can be rewritten as:

$$Q_{k,c} = \frac{2\,\pi\,\lambda_c L(T_{ci} - T_{co})}{\ln (D_{co}/D_{ci})} \tag{2.64}$$

where

- λ_c is the thermal conductivity of the cover;
- T_{co} is the outer cover temperature;
- D_{co} is the outer cover diameter.

λ_c is constant and independent of temperature.

2.3.1.6 External Convective Loss

As seen in Sect. 2.3.1.2, the convective heat transfer between the cover and the environment can be expressed through the Newton's law:

$$Q_{c,ce} = h_{air}\, \pi\, D_{co} L (T_{co} - T_{air}) \tag{2.65}$$

where T_{air} is the ambient temperature. The convective heat transfer coefficient of air is defined as follows:

$$h_{air} = \frac{Nu_{air}\lambda_{air}}{D_{co}} \tag{2.66}$$

where

- Nu_{air} is the Nusselt number for air;
- λ_{air} is the conductive heat transfer coefficient for air, evaluated at the film temperature $(T_{co} + T_{air})/2$.

Convection will be forced or free depending on the presence or absence of wind. If wind is present, heat transfer occurs by forced convection and the following correlation can be employed [15]:

$$Nu_{air} = 0.3 + \frac{0.62\, Re_{air}^{1/2} Pr_{air}^{1/3}}{\left[1 + (0.4/Pr_{air})^{2/3}\right]^{1/4}} \left[1 + \left(\frac{Re_{air}}{282000}\right)^{5/8}\right]^{4/5} \tag{2.67}$$

where

- Re_{air} is the Reynolds number for air evaluated at film temperature $(T_{co} + T_{air})/2$ and characteristic length D_{co};
- Pr_{air} is the Prandtl number for air evaluated at film temperature $(T_{co} + T_{air})/2$.

The correlation can be used for $Re_{air} Pr_{air} > 0.2$.

If there is no wind, the heat transfer between the cover and the environment will be by free convection. In this case, we propose the correlation [16]:

$$Nu_{air} = \left\{0.6 + \frac{0.387\, Ra_{air}^{1/6}}{\left[1 + (0.559/Pr_{air})^{9/16}\right]^{8/27}}\right\}^2 \tag{2.68}$$

This equation considers a long isothermal horizontal cylinder and it can be adopted for $10^5 < Ra_{air} < 10^{12}$. Considerations made for Reynolds and Prandtl numbers in Eq. (2.67) are still valid.

2.3.1.7 External Radiative Loss

The radiative heat transfer between the cover and the environment is caused by the temperature difference between the outer cover surface and the sky. This condition is approximated by considering a small convex gray object (the cover) in a large blackbody cavity (the sky). Therefore, the net exchanged radiation is:

$$Q_{r,ce} = \varepsilon_c \, \pi \, D_{co} L \, \sigma (T_{co}^4 - T_{sky}^4) \tag{2.69}$$

Temperatures are in kelvin and emissivities are constant.

The sky temperature T_{sky} can be related to the dry bulb temperature T_{air} and the dew point ambient temperature T_{dp} as follows [17]:

$$T_{sky} = \varepsilon_{sky}^{1/4} T_{air} \tag{2.70}$$

where the sky emissivity is given by

$$\varepsilon_{sky} = 0.711 + 0.56 \left(\frac{T_{dp}}{100} \right) + 0.73 \left(\frac{T_{dp}}{100} \right)^2 \tag{2.71}$$

2.3.2 Thermal Efficiency

The thermal efficiency of a PTC is defined as the ratio of the useful heat gain of the HTF, Q_u, to the solar energy intercepted by the collector aperture area, S_a, and is given by:

$$\eta = \frac{Q_u}{S_a} = \frac{\dot{m} c_p (T_{fo} - T_{fi})}{G_{bt} A_a} \tag{2.72}$$

where:

- $\dot{m}$ is the mass flow rate of the HTF;
- c_p is the specific heat at constant pressure of the HTF;
- T_{fo} is the outlet fluid temperature;
- T_{fi} is the inlet fluid temperature;
- G_{bt} is the beam radiation measured on the plane of aperture (it must be properly evaluated by using Eqs. (2.14) and (2.15));
- A_a is the collector aperture area.

An energy balance alternative to System (2.48) can be extended to a control volume containing only the absorber. For a PTC of aperture area A_a, the energy balance on the cylindrical absorber yields:

$$S = Q_u + Q_l + \frac{dE_c}{dt} \tag{2.73}$$

where

- S is the solar beam radiation collected in the absorber tube after reflection, defined by Eq. (2.50);
- Q_u is the rate of useful heat gain;
- Q_l is the rate of heat loss from the absorber;
- dE_c/dt is the rate of internal energy storage in the collector.

Equation (2.73) can be rewritten in terms of an overall loss coefficient, U_L, by considering the expression:

$$Q_l = U_L A_r (T_r - T_{air}) \tag{2.74}$$

where

- A_r is the area of the absorber surface (equal to $\pi D_{ao} L$);
- T_r is the average temperature of the absorber surface;
- T_{air} is the ambient temperature.

Substituting Eq. (2.74) in (2.73), operating in steady state conditions ($dE_c/dt = 0$) and rearranging terms, one gets[4]:

$$Q_u = (\tau\alpha)\rho\gamma G_{bt} A_{ae} - U_L A_r (T_r - T_{air}) \tag{2.75}$$

The problem with this equation is that the average temperature of the absorber surface, T_r, is difficult to calculate or measure since it is a function of the collector design, the incident solar radiation and the entering fluid conditions. However, one can use T_{fi}, the inlet fluid temperature, instead of T_r, by introducing a heat removal factor F_R:

$$Q_u = F_R \left[(\tau\alpha)\rho\gamma G_{bt} A_{ae} - U_L A_r (T_{fi} - T_{air}) \right] \tag{2.76}$$

where

$$F_R = \frac{\dot{m} c_p}{A_r U_L} \left[1 - \exp\left(-\frac{A_r U_L F'}{\dot{m} c_p} \right) \right] \tag{2.77}$$

---

[4]Some authors define an effective transmittance-absorptance product which accounts for the reduced thermal losses due to absorption of solar radiation by the cover. However, this effect has been already considered in the energy balance of Eq. (2.48), thus it will be not reconsidered here.

The heat removal factor is an important design parameter since it is a measure of the thermal resistance encountered by the absorbed radiation in reaching the HTF. From Eq. (2.76), it is possible to define F_R as the ratio of the useful heat gain of the fluid to the gain which would occur if the absorber were at temperature T_{fi} everywhere. Note that F_R can range between 0 and 1.

The term F' is the collector efficiency factor and represents the ratio of the useful heat gain of the fluid to the gain which would occur if the whole absorber were at the local fluid temperature. It is given by:

$$F' = \frac{1/U_L}{1/U_L + D_{ao}/(h_f D_{ai}) + D_{ao} \ln(D_{ao}/D_{ai})/(2\lambda_a)} \tag{2.78}$$

where

- h_f is the convective heat transfer coefficient of the fluid defined in Eq. (2.53);
- λ_a is the thermal conductivity of the absorber.

Equation (2.76) is the Hottel–Whillier–Bliss equation adapted for PTCs. Dividing this equation by the solar energy intercepted by the collector aperture area, S_a, one gets:

$$\eta = \frac{Q_u}{S_a} = F_R \left[\eta_o - \frac{U_L}{C} \left(\frac{T_{fi} - T_{air}}{G_{bt}} \right) \right] \tag{2.79}$$

where η_o is the optical efficiency of the PTC, the ratio of solar energy collected by the absorber to that intercepted by the concentrator. The optical efficiency can be written as:

$$\eta_o = (\tau\alpha)\rho\gamma(1 - A_f) \tag{2.80}$$

Equation (2.79) is an alternative form of Eq. (2.72) and allows to determine the thermal efficiency of a PTC as a function of the term $(T_{fi} - T_{air})/G_{bt}$.

As a general comment, it is worth noting that the concentration ratio has a relevant role in reducing the thermal losses of a PTC: from Eq. (2.79), it is evident that the greater is the concentration ratio, the higher is the efficiency. Optical efficiency also assumes a decisive role.

References

1. Duffie JA, Beckman WA (2013) Solar engineering of thermal processes, 4th edn. Wiley, Hoboken
2. Kalogirou SA (2014) Solar energy engineering: processes and systems, 2nd edn. Elsevier, Oxford
3. Rabl A (1976) Comparison of solar concentrators. Sol Energy 18(2):93–111
4. Treadwell GW (1976) Design considerations for parabolic-cylindrical solar collectors. In: Sharing the sun: solar technology in the seventies, vol 2, pp 235–252

5. Guven HM, Bannerot RB (1986) Derivation of universal error parameters for comprehensive optical analysis of parabolic troughs. J Sol Energy Eng 108:275–281
6. Jeter SM, Jarrar DI, Moustafa SA (1983) Geometrical effects on the performance of trough collectors. Sol Energy 30:109–113
7. Coccia G, Latini G, Sotte M (2012) Mathematical modeling of a prototype of parabolic trough solar collector. J Renew Sustain Energy 4(2):023110
8. Forristall R (2003) Heat transfer analysis and modeling of a parabolic trough solar receiver implemented in engineering equation solver. NREL, Golden
9. Sukhatme SP, Nayak JK (2008) Solar energy: principles of thermal collection and storage. Tata McGraw-Hill Publishing Company, New Delhi
10. Gnielinski V (1975) New equations for heat and mass transfer in the turbulent flow in pipes and channels. NASA, STI/Recon technical report A, vol 41, pp 8–16
11. Colebrook CF (1939) Turbulent flow in pipes, with particular reference to the transition region between the smooth and rough pipe laws. J ICE 11:133–156
12. Ratzel A, Hickox C, Gartling D (1979) Techniques for reducing thermal conduction and natural convection heat losses in annular receiver geometries. J Heat Trans-T ASME 101(1):108–113
13. Hollands KGT, Raithby GD, Lonicek L (1975) Correlation equations for free convection heat transfer in horizontal layers of air and water. Int J Heat Mass Transf 18(7):879–884
14. Marshal N (1976) Gas encyclopedia. Elsevier, New York
15. Churchill SW, Bernstein M (1977) A correlating equation for forced convection from gases and liquids to a circular cylinder in crossflow. J Heat Trans-T ASME 99:300–306
16. Churchill SW, Chu HHS (1975) Correlating equations for laminar and turbulent free convection from a horizontal cylinder. Int J Heat Mass Transf 18(9):1049–1053
17. Martin M, Berdahl P (1984) Characteristics of infrared sky radiation in the United States. Sol Energy 33(3):321–336

Chapter 3
Standards and Testing

Abstract The performance of solar thermal collectors such as PTCs can be assessed by performing specific procedures described in standards. Standards generally followed in the solar energy field are the ISO 9806, the ANSI/ASHRAE 93 and the EN 12975-1, which will be presented in this chapter. The measurements and the procedures required for testing PTCs will be discussed in detail, in particular focusing the attention on the three most important parameters of a solar collector: the time constant, the thermal efficiency and the incident angle modifier. Due to its importance, uncertainty in thermal efficiency testing will be described extensively. Also, quality test methods will be briefly discussed.

Keywords Uncertainty · Thermal efficiency · Incidence angle modifier · Time constant · Quality test

3.1 Available Standards

Different standards can be adopted to evaluate the thermal performances of concentrating solar collectors such as PTCs. In this section, the international, US and European standards dedicated to this purpose are presented. To date, the standards which should be considered are the ISO 9806:2013 and the ANSI/ASHRAE Standard 93–2010 (RA2014).

3.1.1 International Standards (ISO)

The ISO 9806:2013 [1] defines procedures for testing fluid heating solar collectors for performance, reliability, durability and safety under well-defined and repeatable conditions. It contains performance test methods for conducting tests outdoors under natural solar irradiance and natural/simulated wind and for conducting tests indoors under simulated solar irradiance and wind.

© The Author(s) 2016
G. Coccia et al., *Parabolic Trough Collector Prototypes for Low-Temperature Process Heat*, SpringerBriefs in Applied Sciences and Technology,
DOI 10.1007/978-3-319-27084-5_3

The standard includes test methods for the steady-state and quasi-dynamic thermal performance of glazed and unglazed liquid heating solar collectors and steady-state thermal performance of glazed and unglazed air heating solar collectors. Collectors tested according to this standard represent a wide range of applications, including tracking concentrating collectors for process heat at low temperature.

The standard is not applicable to those collectors in which the thermal storage unit is an integral part of the collector to such an extent that the collection process cannot be separated from the storage process for the purpose of making measurements of these two processes.

The ISO 9806:2013 cancels and replaces the ISO 9806-1:1994, ISO 9806-2:1995 and ISO 9806-3:1995, which have been technically revised. It also replaces the EN 12975-2:2006 (see Sect. 3.1.3).

3.1.2 US Standards

The ANSI/ASHRAE Standard 93–2010 (RA2014) [2] provides a test procedure for solar energy collectors which use single-phase fluids (liquids or gases, but not a mixture of the two phases) and have no significant internal energy storage. Collectors can be tested both indoors (under a simulated solar irradiance) and outdoors (under natural solar irradiance) to determine steady and quasi-steady[1] state thermal performance, time constants and variations in efficiency with changes in the angle of incidence.

The standard defines its applicability to both liquid-cooled concentrating and non-concentrating collectors, and air collectors. It does not apply to unglazed solar collectors and those collectors in which the heat transfer fluid changes phase and the leaving transfer fluid contains vapor.

First published in 1986 [3], the standard was reaffirmed in 1991, in 2003 and again in 2010.

3.1.3 European Standards

The first two European standards adopted for testing solar collectors were the EN 12975-1:2000 [4] and the EN 12975-2:2001 [5]. The former specified requirements on durability, reliability and safety for liquid heating solar collectors, while the latter specified test methods to validate those requirements and included three test methods for the thermal performance characterization for liquid heating collectors. Both the

---

[1]Quasi-steady state describes solar collector test conditions when the flow rate, inlet fluid temperature, collector temperature, solar irradiance, and ambient environment have stabilized such an extent that these conditions may be considered essentially constant.

standards were not applicable to those collectors in which the thermal storage unit is an integral part of the collector and to tracking concentrating solar collectors.[2]

The EN 12975-1 and 12975-2 were reaffirmed in 2006. In 2010, the amendment A1:2010 was added to the EN 12975-1 to include concentrating solar collectors. As noted in Sect. 3.1.1, the EN 19275-2:2006 has been replaced by the ISO 9806:2013.

3.2 Performance Test Computations

The performance of a PTC can be determined by calculating the instantaneous efficiency for different values of:

- the inlet fluid temperature;
- the ambient temperature;
- the incident solar radiation.

This requires to measure experimentally, under steady state or quasi-steady state conditions:

- the mass flow rate of the HTF, $\dot{m}$;
- the inlet fluid temperature, T_{fi};
- the outlet fluid temperature, T_{fo};
- the ambient air temperature, T_{air};
- the beam solar radiation on the plane of aperture, G_{bt};
- the wind velocity, w_{air}.

The aperture area, A_a, and the specific heat capacity of the HTF, c_p, should also be measured with certain accuracy, since they appear in the expression of the thermal efficiency reported in Eq. (2.72).

3.3 Measurement Requirements

The following measurements are the most important required by the ISO 9806:2013 to test liquid concentrating collectors glazed.

- The accuracy of the liquid flow rate measurements should be equal to or better than $\pm 1.0\%$ of the measured value in mass per unit time. The mass flow rate of the HTF should be the same ($\pm 10\%$) throughout the whole test sequence.
- Temperature measurements could require different accuracies, thus different sensors can be used (e.g., thermocouples, thermopiles and resistance thermal detectors). In particular:

---

[2]EN 12975-2:2001 specified that, even though basically not applicable to tracking concentrating collectors, given quasi-dynamic testing was also applicable to most concentrating collector designs.

– the inlet fluid temperature should be measured to a standard uncertainty of
  0.1 K but, in order to check that the temperature is not drifting with time, a
  better resolution of the temperature to $+0.02$ K could be required;
– the difference between the outlet and inlet fluid temperatures should be determined to a standard uncertainty lesser than 0.05 K;
– the ambient temperature should be measured to a standard uncertainty lesser
  than 0.5 K.

• The inlet pressure and the pressure drop across the collector should be measured
  with a device having an error lesser than 5% of the measured value or ± 10 Pa.
• Pyranometers and pyrheliometers can be used for measuring solar radiation. These
  instruments should have minimum characteristics defined by the standards in detail
  and should be calibrated for solar response within one year preceding the collector
  tests against another pyranometer whose calibration uncertainty relative to recog-
  nized measurement standards is known. Recall that, as discussed in Sect. 2.1.4,
  pyranometers measure the global solar radiation on the horizontal plane, while
  pyrheliometers detect the direct solar radiation on a plane normal to the radiation
  itself. The combination of these two instruments generally delivers more accurate
  results.
• The average wind velocity should be measured to a standard uncertainty lesser than
  0.5 m/s (0.25 m/s for unglazed collectors) for both indoor and outdoor testing. The
  measurement of the average value may be obtained either by an arithmetic average
  of sampled values or by a time integration over the test period.
• The collector gross area should be measured to a standard uncertainty of 0.3%.
• The specific heat capacity c_p and the density ρ of the fluid should be known to
  within $\pm 1\%$ over the range of fluid temperatures used during the tests.

The following correlations can be used for water at 1 bar and at a temperature between
0 and 99.5 °C. For the specific heat capacity:

$$c_p = 4.217 - 3.358 \times 10^{-3}T + 1.089 \times 10^{-4}T - 1.675 \times 10^{-6}T$$
$$+ 1.309 \times 10^{-8}T - 3.884 \times 10^{-11}T \tag{3.1}$$

And for the density:

$$\rho = 999.85 + 6.187 \times 10^{-2}T - 7.654 \times 10^{-3}T + 3.974 \times 10^{-5}T$$
$$- 1.110 \times 10^{-7}T \tag{3.2}$$

Acceptable test configurations for testing liquid solar collectors are:

• closed-loop;
• open-loop;
• open-loop with fluid continuously supplied.

All these configurations are acceptable if the specified test conditions are satisfied.
Figure 3.1 shows an example of closed loop testing configuration when the heat
transfer fluid is a liquid.

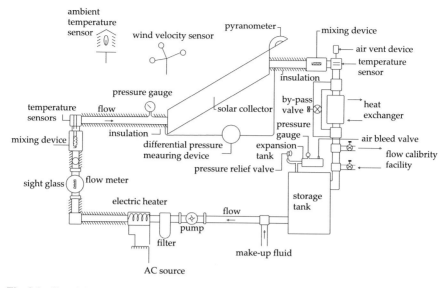

Fig. 3.1 Closed-loop testing configuration for liquid solar collectors. Adapted from [2]

3.4 PTC Parameters

Tests are performed to determine the time response characteristics of the collector as well as how its steady-state thermal efficiency varies with the angle of incidence at various Sun and collector positions. The parameters described in the following were developed to control test conditions so that a well-defined efficiency curve can be obtained with a minimum of data scatter.

3.4.1 Time Constant

The determination of the time response of a solar collector is required to evaluate the transient behavior of the collector itself and to select the proper time intervals for the quasi-steady or steady-state efficiency tests. Whenever transient conditions exist, Eq. (2.72) does not govern the thermal performance of the collector, since part of the absorbed solar energy is used for heating up the collector or, if energy is lost, it results in cooling the collector. In this case, the transient behavior of a solar collector is described by Eq. (2.73), and the following assumptions are considered valid:

- The beam radiation G_{bt} is initially zero and is suddenly increased and held constant.
- η_o, U_L, T_{air}, $\dot{m}$, and c_p are considered constant during the transient period.
- The rate of change of the mean fluid temperature, T_{fm}, with time is related to the rate of change of the outlet fluid temperature, T_{fo}, with time by

$$\frac{\mathrm{d}T_{\mathrm{fm}}}{\mathrm{d}t} = K\frac{\mathrm{d}T_{\mathrm{fo}}}{\mathrm{d}t} \tag{3.3}$$

where K is a dimensionless capacity factor which affects the rate at which a particular collector will heat up to steady state and to a stabilized outlet temperature, $T_{\mathrm{fo,s}}$. K is defined as

$$K = \frac{\dot{m}c_p}{F'A_a U_L}\left(\frac{F'}{F_R} - 1\right) \tag{3.4}$$

With these assumptions, the collector time constant is defined as the time, t, the collector takes to go from its delta temperature with no solar radiation to 63.2% of its delta temperature at a steady state condition, given a stable exposure:

$$\frac{T_{\mathrm{fo}}(t) - T_{\mathrm{fi}}}{T_{\mathrm{fo,s}} - T_{\mathrm{fi}}} = 0.632 \tag{3.5}$$

3.4.2 Thermal Efficiency

The thermal efficiency of a PTC is given by Eqs. (2.72) and (2.79):

$$\eta = \frac{\dot{m}c_p(T_{\mathrm{fo}} - T_{\mathrm{fi}})}{G_{\mathrm{bt}}A_a} = F_R\left[\eta_o - \frac{U_L}{C}\left(\frac{T_{\mathrm{fi}} - T_{\mathrm{air}}}{G_{\mathrm{bt}}}\right)\right] \tag{3.6}$$

Equation (3.6) states that if the efficiency, η, is plotted as a function of $(T_{\mathrm{fi}} - T_{\mathrm{air}})/G_{\mathrm{bt}}$, a straight line will result considering U_L as a constant. Note that F_R is dependent on U_L, as Eq. (2.77) shows. This straight line has the following parameters:

- the intercept is equal to $F_R\eta_o$;
- the slope is equal to $-(F_R U_L)/C$.

The thermal efficiency reaches a maximum (i.e., the intercept value) when the inlet fluid temperature equals the ambient temperature. On the contrary, the thermal efficiency is zero when the radiation level is low or the fluid temperature high so that heat losses equal solar absorption and there is no useful heat gain. This last condition is called stagnation and usually occurs when no fluid flows in the collector. Collectors must be designed to withstand stagnation temperatures. The maximum temperature is given by:

$$T_{\mathrm{f,max}} = T_{\mathrm{air}} + \frac{\eta_o C G_{\mathrm{bt}}}{U_L} \tag{3.7}$$

It is worth noting that U_L is not always a constant but may be a function of the temperature of the absorber and of the weather conditions. Similarly, the optical efficiency, η_o, varies with the angle of incidence.

3.4.3   Incident Angle Modifier

The optical efficiency in Eq. (3.6) is dependent on the angle of incidence and for off-normal incidence angles is difficult to be described analytically and measured, since it strongly depends on the collector geometry and optics. However, the actual optical efficiency can be replaced by its value at normal incidence, $\eta_{o,n}$, if a factor called incident angle modifier, $K_{\tau\alpha}$, is provided. The incident angle modifier is given by:

$$K_{\tau\alpha} = \frac{(\tau\alpha)\rho\gamma(1 - A_f)}{[(\tau\alpha)\rho\gamma]_n(1 - A_{f,n})} = \frac{\eta_o}{\eta_{o,n}} \qquad (3.8)$$

therefore Eq. (3.6) becomes

$$\eta = F_R \left[K_{\tau\alpha}\eta_{o,n} - \frac{U_L}{C}\left(\frac{T_{fi} - T_{air}}{G_{bt}} \right) \right] \qquad (3.9)$$

Using the incident angle modifier, the thermal efficiency of a PTC can be determined at normal or near-normal incidence conditions. In this way, $K_{\tau\alpha} \simeq 1$ and the intercept of the efficiency curve is equal to $F_R\eta_{o,n}$. A separate measurement is conduced to determine the value of $K_{\tau\alpha}$ so that the collector performances can be predicted under a wide range of conditions and/or time of day using Eq. (3.9).

3.5   Performance Test Procedures

To correctly characterize a glazed PTC, outdoor tests should be performed by satisfying the following conditions.

- The fluid flow rate should be set at approximately 0.02 kg/s per square meter of collector gross area. It should be held stable to within $\pm 2\%$ of the set value during each test period, and should not vary by more than $\pm 10\%$ of the set value from one test period to another. In any case, transitional regimes should be avoided.
- The hemispherical solar irradiance at the plane of the collector aperture should be greater than $700\,W/m^2$.
- The angle of incidence of direct solar radiation at the collector aperture should be in the range in which the incident angle modifier for the collector varies by no more than $\pm 2\%$ from its value at normal incidence. In order to characterize the PTC performance at various angles, an incident angle modifier must be determined (Sect. 3.4.3).
- The average value of wind velocity parallel to the collector aperture should be 3 ± 1 m/s.
- The inlet fluid temperature should remain constant within the operating range. In fact, small variations in the inlet temperature could lead to errors in the estimated thermal efficiency.

- Data points should be obtained for at least four different inlet fluid temperatures spaced evenly over the operating temperature range of the collector. A minimum of four independent data points should be registered for each inlet fluid temperature, for a total of sixteen data points. It is important that one inlet temperature is selected such that it lies within ± 3 K of the ambient temperature, in order to obtain an accurate determination of the incident angle modifier.
- The ambient temperature should be lesser than 30 °C.

If the previous conditions are satisfied, the following testing procedure allows to determine the parameters which characterize the performance of a PTC, as described in Sect. 3.4.

1. At first, a performance test is conducted on the PTC to determine its time constant. The inlet fluid temperature is adjusted as closely as possible to the ambient temperature and is controlled while circulating the HTF through the collector at the specified flow rate and maintaining steady or quasi-steady state conditions with the collector covered. Then, the incident solar radiation is abruptly increased to a value greater than 700 W/m². Inlet and outlet fluid temperatures must be continuously monitored as functions of time until a steady state condition is achieved, that is when:

$$\Delta T_{\text{s}} = T_{\text{fo,s}} - T_{\text{fi}} \tag{3.10}$$

The actual time constant is the time t required to the collector to reach the condition expressed in Eq. (3.5).

2. When the first test is completed, a series of thermal efficiency tests are conducted at near-normal incident conditions. The angle of incidence should be in the range in which the incident angle modifier varies by no more than $\pm 2\%$ from the normal incidence value. An acceptable evenly spaced distribution of inlet temperatures can be obtained by setting $(T_{\text{fi}} - T_{\text{air}})$ to 0, 30, 60 and 90% of the value of $(T_{\text{fi}} - T_{\text{air}})$ reached at the maximum tested inlet temperature and at a given ambient temperature. At least four data points should be taken for each value of T_{fi} at steady or quasi-steady state conditions. The ambient temperature should not vary by more than ± 1.5 °C.

3. Finally, the PTC incident angle modifier is determined as a function of the angle of incidence. The orientation of the collector should be such that the collector is maintained within $\pm 2.5°$ of the angle of incidence for which the test is being conducted. For PTCs, the collector should be oriented so that the test incident angles are, approximately, 0, 30, 45, and 60°. It is recommended that these data be taken during a single day. For each data point, the inlet temperature should be as closely as possible (± 1 °C) to the ambient temperature, so that from Eq. (3.9) one gets:

$$K_{\tau\alpha} \simeq \frac{\eta}{F_{\text{R}}\eta_{0,\text{n}}} \tag{3.11}$$

Since $F_R \eta_{0,n}$ will have already been obtained as the intercept of the efficiency curve, the values of $K_{\tau\alpha}$ can be computed for different angles of incidence with Eq. (3.11). An expression for the incident angle modifier, i.e. $K_{\tau\alpha}(\theta)$, can be obtained by using curve fitting methods.

3.6 Uncertainty in Solar Collector Thermal Efficiency Testing

One of the aims of the performance tests is to determine the thermal efficiency of a PTC. But this parameter is the result of a series of measurements, thus it is only an approximation of the real value since all measurements are affected to uncertainty. The repetition of a measurement does not guarantee that the obtained results are always the same; instead, it is usually verified that the results of a repeated measurement are included in a well-defined range of values.

The result of a measurement should be expressed in a reliable form to be used in a useful way. In fact, without such form, the results of a measurement cannot be compared one another. This form can be expressed in terms of uncertainty of a measurement. Uncertainty in measurement is defined by international standards. One of these standards is the GUM, "Guide to the expression of Uncertainty in Measurement" [6], to which we will refer to develop an analysis of the uncertainty in PTC thermal efficiency testing.

As discussed in the previous sections, the thermal efficiency η cannot be measured directly, since it depends on N other input quantities through the following relationship:

$$\eta = f(\rho, \dot{V}, c_p, \Delta T, G_{bn}, \theta, A_a) = \frac{\rho \dot{V} c_p \Delta T}{G_{bn} \cos \theta A_a} \qquad (3.12)$$

where ρ is the fluid density, $\dot{V}$ is the volumetric flow rate and ΔT is the temperature difference between the outlet and the inlet of the receiver.

The best estimate (or expectation) of η can be written as:

$$\bar{\eta} = f(\bar{\rho}, \overline{\dot{V}}, \overline{c_p}, \overline{\Delta T}, \overline{G_{bn}}, \bar{\theta}, \overline{A_a}) = \frac{\bar{\rho} \overline{\dot{V}} \, \overline{c_p} \, \overline{\Delta T}}{\overline{G_{bn}} \cos \bar{\theta} \, \overline{A_a}} \qquad (3.13)$$

where the bar indicates the best estimate of each quantity. The following sections describe how to calculate the aforementioned estimates, their uncertainties and the global uncertainty of the estimate of the thermal efficiency.

3.6.1 Type A and B Uncertainties of the Input Quantities

According to the recommendations of the GUM, uncertainties are classified in two types:

- Type A uncertainties are evaluated by means of a statistical analysis of series of observations;
- Type B uncertainties are evaluated by methods different from the statistical analysis.

The information used to estimate Type A uncertainty derives from the experiment/measurement being studied, while Type B uncertainty derives from external sources, e.g. previous measurements, experience or general knowledge of the properties of the used materials/instruments, data declared by the manufacturer, etc.

When an input quantity X is obtained experimentally by repeated measurements, uncertainty must be evaluated according to the Type A approach. Let us consider N statistically independent observations x_i of X; the best estimate of X is the arithmetic mean (or mean) of the N observations x_i:

$$\bar{x} = \frac{1}{N} \sum_{i=1}^{N} x_i \qquad (3.14)$$

The best estimate of $\bar{x}$ is its standard deviation $\sigma(\bar{x})$, called experimental standard deviation of the mean and given by:

$$\sigma(\bar{x}) = \sqrt{\frac{1}{N(N-1)} \sum_{i=1}^{N} (x_i - \bar{x})^2} \qquad (3.15)$$

The experimental standard deviation of the mean indicates how well $\bar{x}$ estimates the expectation of X. In other words, $\sigma(\bar{x})$ can be considered as a measure of the uncertainty of $\bar{x}$ and is defined Type A standard uncertainty:

$$u_{\mathrm{A}}(\bar{x}) = \sigma(\bar{x}) \qquad (3.16)$$

As concerns the thermal efficiency of a PTC, u_{A} should be generally calculated for $\overline{V}$, $\overline{\Delta T}$ and $\overline{G_{\mathrm{bn}}}$.

For an estimate $\bar{x}$ of an input quantity X which has not been obtained from repeated observations, the associated estimated uncertainty is evaluated by scientific judgment based on all of the available information on the possible variability of X. For example, when it is possible to know the upper (a^+) and lower (a^-) bounds assumed by X and there is no specific knowledge about the possible values of X within the interval, one can assume that it is equally probable for X to lie anywhere within it; in other words, a uniform (or rectangular) distribution of possible values can be considered. In this case, the Type B standard uncertainty associated to the expectation of X is:

$$u_B(\bar{x}) = \sqrt{\frac{(a^+ - a^-)^2}{12}} \qquad (3.17)$$

Finally, the Type A and B uncertainties of the estimate of an input quantity X can be combined together through the following expression:

$$u(\bar{x}) = \sqrt{u_A^2(\bar{x}) + u_B^2(\bar{x})} \qquad (3.18)$$

where $u(\bar{x})$ denotes the combined uncertainty of the estimate of the input quantity.

3.6.2 Law of Propagation of Uncertainty

The law of propagation of uncertainties allows to calculate the uncertainty of the estimate of an output quantity Y when the uncertainty of the estimates of its input quantities X_i are known. Referring to the purpose of the present work, the law allows to calculate the combined uncertainty of $\bar{\eta}$ given the combined uncertainties of the estimates of its input quantities (see Eq. 3.13). It is necessary to consider two different input quantities:

- independent (or uncorrelated);
- interdependent (or correlated).

If all the input quantities are independent, the combined standard uncertainty of the estimate of the output is given by:

$$u_{ind}(\bar{y}) = \sqrt{\sum_{i=1}^{N} \left(\frac{\partial f}{\partial \bar{x}_i} \right) u^2(\bar{x}_i)} \qquad (3.19)$$

Considering for f the expression given in Eq. 3.13, the law of propagation of uncertainty for the thermal efficiency can be written as:

$$\frac{u_{ind}(\bar{\eta})}{\bar{\eta}} = \left[\left(\frac{u(\bar{\rho})}{\bar{\rho}} \right)^2 + \left(\frac{u(\bar{V})}{\bar{V}} \right)^2 + \left(\frac{u(\bar{c}_p)}{\bar{c}_p} \right)^2 + \left(\frac{u(\overline{\Delta T})}{\overline{\Delta T}} \right)^2 \right.$$
$$\left. + \left(\frac{u(\overline{G_{bn}})}{\overline{G}_{bn}} \right)^2 + \left(\frac{u(\bar{\theta})}{\bar{\theta}} \right)^2 + \left(\frac{u(\overline{A_a})}{\overline{A}_a} \right)^2 \right]^{\frac{1}{2}} \qquad (3.20)$$

Otherwise, if among two or more input quantities exist a certain degree of correlation (e.g., the quantities are estimated with the same instrument), it is necessary to consider the more general case of law of propagation of uncertainty which can be expressed as:

$$u_{\mathrm{dep}}(\bar{y}) = \sqrt{\sum_{i=1}^{N} \sum_{j=1}^{N} \frac{\partial f}{\partial \bar{x}_i} \frac{\partial f}{\partial \bar{x}_j} u(\bar{x}_i, \bar{x}_j)} \qquad (3.21)$$

where $u(\bar{x}_i, \bar{x}_j)$ is the estimated covariance, a parameter that exhibits the degree of statistical dependence among the estimates of two input quantities:

$$u(\bar{x}_i, \bar{x}_j) = \frac{1}{N(N-1)} \sum_{i,j=1}^{N} (x_i - \bar{x}_i)(x_j - \bar{x}_j) \qquad (3.22)$$

3.7 Quality Test Methods

In addition to tests which provide the performance characteristics of a PTC, it is also important to conduce tests to determine the quality of a collector. Therefore, all the available standards include quality tests necessary to verify the resistance of a solar collector to different environmental conditions.

These tests are briefly described in the following list, valid for liquid non-organic solar collectors. The tests should be carried out in the reported sequence. It is suggested to consult one of the standards discussed in Sect. 3.1 if further information is required.

1. Internal pressure test: the fluid channels should be tested to verify the extent to which they can withstand the pressures which they might meet in service.
2. High-temperature resistance test: a collector should withstand high temperature and irradiance levels without failures (e.g., glass breakage, collapse of plastic cover, melting of plastic absorber, etc.).
3. Standard stagnation temperature.
4. Exposure and pre-exposure test: it provides a low-cost reliability test sequence indicating (or simulating) operating conditions which are likely to occur during real service.
5. External thermal shock test: it assesses the capability of a collector to withstand thermal shocks such as rainstorms or hot sunny days without a failure.
6. Internal thermal shock test: to check the capability of the collector to withstand a cold HTF on hot sunny days without a failure.
7. Rain penetration test: applicable only for glazed collectors, it assess the extent to which glazed collectors are resistant to rain penetration.
8. Freeze resistance test.
9. Mechanical load test with positive and negative pressure: the former is necessary to check if the transparent cover of the collector, the collector box and the fixings are able to resist the positive pressure load due to the effect of wind and snow, while the latter to assess the deformation and the extent to which the collector box and the fixings are able to resist uplift forces caused by the wind.
10 Impact resistance test: to verify the resistance of a collector to the effect of impacts caused by hailstones.

References

1. ISO 9806:2013 (2013) Solar energy—solar thermal collectors—test methods
2. ANSI/ASHRAE Standard 93–2010 (RA2014) (2014) Methods of testing to determine the thermal performance of solar collectors
3. ANSI/ASHRAE Standard 93–1986 (1986) Methods of testing to determine the thermal performance of solar collectors
4. EN 12975–1:2000 (2000) Thermal solar systems and components—solar collectors—part 1: general requirements
5. EN 12975–2:2001 (2001) Thermal solar systems and components—solar collectors—part 2: test methods
6. JCGM 100:2008 (2008) Evaluation of measurement data. Guide to the expression of uncertainty in measurement

Chapter 4
Concentrator

Abstract PTCs are solar collectors made by several components, nonetheless they can be divided into two different parts which have distinct functions: the concentrator and the receiver. The former is the subject of the present chapter. The scope of PTC concentrators is to reflect the maximum amount of solar radiation to the receiver. In PTCs, concentrators have the shape of a cylindric parabola, a structure which has the property to reflect each normal incident solar ray to a line belonging to the parabola itself and called focal line, where the receiver is located. In order to correctly concentrate the solar radiation on the receiver, it is crucial to obtain a concentrator with good characteristics. This can be achieved by designing an accurate mold and by choosing appropriate materials for both the concentrator structure itself and the reflective foil attached to it. This chapter analyzes all the aforementioned aspects by discussing the manufacture of several concentrators of PTC prototypes available in literature. An overview on adopted materials is also presented.

Keywords Parabola · Mold · Accuracy · Fiberglass · Aluminum

4.1 Introduction to PTC Concentrators

As discussed in Sect. 1.2, there are several types of concentrators used in concentrating solar collectors. Concentrators allow to concentrate the sun rays to a specific point, as in the heliostat field collectors, where several mirrors direct the rays to a single point. There are also other kinds of concentrators which concentrate the solar radiation onto a tubular surface, as required by PTCs.

The concentrator of a solar collector is an optical device able to reflect the sun rays to the receiver thanks to the reflecting materials by which it is built (Fig. 4.1). These materials are usually optical mirrors with high reflection grade. The concentrator has a parabolic form to concentrate each sun ray in a line, which corresponds to the parabola focus.

© The Author(s) 2016
G. Coccia et al., *Parabolic Trough Collector Prototypes for Low-Temperature Process Heat*, SpringerBriefs in Applied Sciences and Technology,
DOI 10.1007/978-3-319-27084-5_4

Fig. 4.1 Schematic of a
PTC concentrator with the
receiver

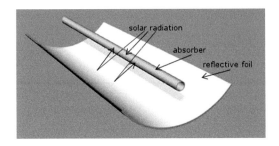

The surface of the collector exposed to the solar rays is called aperture area, while the surface where the rays are focused is named receiver area. These two areas are related one another through the concentration ratio C, defined in Eq. (2.16):

$$C = \frac{A_a}{A_r} \qquad (4.1)$$

It is worth noting that, due to the presence of the receiver, there is portion of the aperture area which is not reached by the sun rays, as shown in Fig. 4.1. This effect was discussed in Sect. 2.2.4.

4.2 Manufacture of a PTC Concentrator

The accuracy of the parabola design is directly related to the accuracy of the mold. The mold is generally made using plywood with different thickness. To ensure the proper accuracy, it is necessary to wait between one step and another, in order to treat any type of material added and to fix every imperfection produced as a consequence of the nicked wood or the scratched surfaces during the building process. Besides, it is common to add a topcoat to reach the final desired quality. The topcoat needs a cure time to prevent troubles further on.

The following sections report the characteristics of PTC concentrators developed as prototypes. For the sake of simplicity, each section has been named with the name of the authors involved in the manufacture. Table 4.1 summarizes the characteristics of the concentrators being discussed.

4.2.1 Kalogirou et al. (1994)

To make the mold of their solar collector, Kalogirou et al. [1] utilized synthetic wood with medium density fibers (MDF) and 18 mm thickness. Basically, the main

Table 4.1 Characteristics of the different concentrators

Description	[1]	[2]	[3]	[4] PTC90	[4] PTC45	[5]	[6]	[7]
Rim angle (°)	90	82.2	90	90	45	100	90	90
Parabola length (m)	2.4	5	1.25	–	–	3	2	2
Parabola width (m)	1.46	1.5	0.8	1.063	1.187	1.67	0.925	0.7
Focal distance (m)	0.365	–	0.2	0.266	0.766	0.35	0.25	0.175
Aperture area (m^2)	3.5	–	1	3	3	11.79	1.85	1.4
Thickness (mm)	8	–	7	–	–	–	–	0.8
Concentration ratio	21.2	16.7	19.89	13.3	14.9	21	9.25	–
Mirror specular reflectance	–	0.83	0.97	0.92	0.92	0.68	0.94	–
Intercept factor	0.98	0.823	0.879	0.84	0.58	–	0.829	–

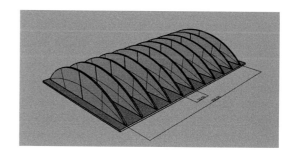

Fig. 4.2 The mold used by Kalogirou et al. (1994)

idea was to make a mold structure formed by 12 pieces; one as a base and 11 with a parabolic shape screwed perpendicularly to the former and separated one from another by 0.20 m (Fig. 4.2). In order to guarantee a precise disposition, a highly accurate aluminum bar was utilized. The first parabolic piece was made manually, following a piece of tracing paper carefully. Once realized this piece, it was used as a guide master to obtain the rest of the pieces utilizing a vertical spindle moulder machine.

On top of the parabolic pieces, a 4 mm-thick melamine plate was nailed. Finally, to avoid the imperfections caused by the nails and to ensure a homogeneous surface, stucco was added and, after a reasonable time, sandpapered. Once stucco was dried and properly fixed, the mold was ready to be used.

The final parabolic trough was made of fiberglass, which is a reflective material. Before covering the wooden mold with fiberglass, a thin layer of a waxy parting

Fig. 4.3 Design of the
concentrator realized by
Kalogirou et al. (1994)

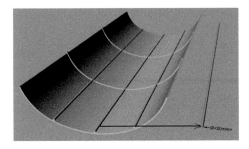

Fig. 4.4 Rib used by Brooks
et al. (2005)

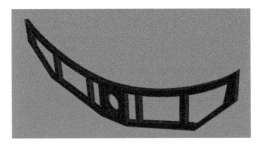

compound was spread along the whole surface to facilitate the later removal. Then,
two additional layers of a polyester resin and woven fiberglass were added. Between
both, plastic conduits were also disposed longitudinally and perpendicularly, forming
perfectly ordered rectangles to stiffen the molded layer (see Fig. 4.3).

4.2.2   Brooks et al. (2005)

Polypropylene and aluminum are the materials used to realize the mold of the present
concentrator [2]. The former was used to make the parabolic ribs, which consist in
seven 25 mm thick rings cut with a computer high-pressure abrasive water-jet process.
Each ring, as shown in Fig. 4.4, is complex, since it contains several spaces inside.
The aluminum, on the other hand, was used to fabricate a tube employed to pierce
each ring through a hole located in its center. The outer diameter of the aluminum
tube is 101.6 mm, with a thickness of 6.4 mm. The separation between each ring was
804.16 mm. The length of the whole structure was 5 m (Fig. 4.5).

In this concentrator, the outer layers and the reflective surface were easily inter-
changeable with the intention of testing different reflective materials. In particular, it
was firstly used a SA-85 aluminum acrylic film as interim material. The final material was vapor-deposited aluminum on a stainless steel substrate, which is in direct
contact with the rings.

Fig. 4.5 Mold by Brooks et al. (2005)

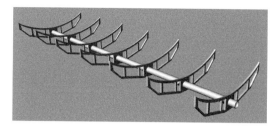

Fig. 4.6 Mold used by Valan Arasu and Sornakumar (2007)

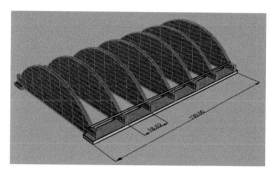

4.2.3 Valan Arasu and Sornakumar (2007)

In the Valan Arasu and Sornakumar's project [3], the wood used to make the mold (plywood ISI BWR-IS-303) was specifically selected to possess high density. To make the plates with a parabolic shape, the design was firstly realized with AutoCAD and then stuck on one of the plates to be crafted closely to the drawing line by a vertical saw machine. To reach the necessary precision during cutting, the parabola Eq. (2.19) was used.

The piece previously modeled was utilized as a reference to cut the other six. Each one of them was 19 mm thick and there was 186.2 mm between each other, conferring homogeneous separation and stability. They were positioned vertically and perpendicularly on two parallel plates made of the same material but with 90 mm width and 1300 mm length separated by 890 mm, as shown in Fig. 4.6. This alignment was achieved doing two small recesses which were then nailed. In order to join all the plates precisely, a rib was adjusted in both ends. The precision in the position and in the verticalness in this step is very important to obtain a good concentrator mirror.

On the parabolic pieces, a 4 mm thick plywood smooth surface was nailed. Finally, the imperfections produced during the manufacture process, such as scratches or marks, were eliminated with wood putty. Once finished the mold and dried the wood putty, two layers were fixed before introducing the layer of polyester resin and chopped strand fiberglass. The first layer consisted of wax polish to separate the casting from the mold. The second one, on the other hand, was polyvinyl alcohol,

Fig. 4.7 Parabola
manufactured by Valan
Arasu and Sornakumar
(2007)

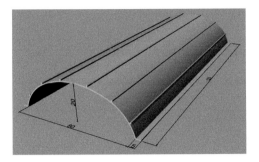

which is very important as it is placed on the reflective surface. During these steps, it is important to respect the curing times.

Then, a 7 mm thick layer of polyester resin and chopped strand fiberglass was introduced. This layer was separated in two other layers with different thickness (3 and 6 mm). Between them, six conduits of polyvinyl chloride (PVC) was longitudinally positioned to reinforce the structure. According to different wind loads studies, the longitudinal disposition resulted strong enough to resist aerodynamic forces. This saves material and reduces weigh considerably. The final structure can be observed in Fig. 4.7.

The authors of this PTC also carried out two load tests: a gravity force and a wind load test. The latter generates more torque load than the gravity force and was evaluating using a method reported in a standard. The tests were conduced using sand as material to simulate the load. The sand was leveled uniformly over the entire trough and the deflection was measured by a dial gauge. The highest velocity estimated in the test region (Madurai City) was 34 m/s, which corresponds to 72 kg of wind load. The measured deflection is 0.95 mm, which is well within acceptable limits.

4.2.4 Jaramillo et al. (2013)

In this project [4], two prototypes were designed, PTC90 and PTC45, with an opening angle of 90° and 45°, respectively. The only different step of the construction process was the rim angle.

To construct the basal parabolic structure, aluminum rings and tubes were utilized instead of wood. The rings allowed to reach the desirable parabolic shape and the tubes, on the other hand, were used to place them on the correct location. These rings, as reported in Fig. 4.8, have to be homogeneously and perpendicularly distributed with respect to the aforementioned tubes, leaving the same distance between them.

The thickness of the rings was around 10 mm and they were made using a numerically control machine, while the external diameter of the tubes was 25.4 mm. The whole mirror structure is shown in Fig. 4.9.

Fig. 4.8 Aluminum ring used in the collector by Jaramillo et al. (2013)

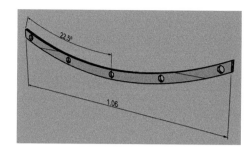

Fig. 4.9 Mirror structure by Jaramillo et al. (2013)

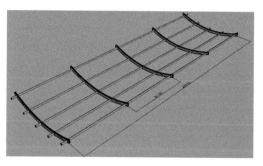

To obtain the definitive parabolic through, also made of aluminum (4270AG), the aluminum foils were directly positioned on the parabolic structure avoiding cuts or any other machining process. By resting on the structure, its own weigh made these reflective plates to be bended, obtaining the required parabolic form. Once finalized both parabolic troughs, they were tested to determine their resistance against static loads. As they are usually located outside, one of the most determinant factor is the effect of wind. Thus, different tests were carried out, measuring the effect of 200 N and 300 N forces (which correspond to 26 km/h and 33 km/h, respectively). With the aim of relating the wind velocity to the applied load, the following Eq. (4.2) was considered:

$$F_W = \frac{1}{2}\rho_a C_D A_a V_w^2 \qquad (4.2)$$

where F_W is the applied force, ρ_a is the density of air, C_D is the drag coefficient and V_W is the wind velocity.

It is important to mention that the major registered force in the area where these types of solar collectors could be installed was around 33 km/h. Firstly, a 200 N force was applied on each corner of the collector, the most susceptible regions to be bent according to the finite elements method used by the authors. Subsequently, two 200 N forces and two 300 N forces were applied to the internal ribs. Finally, four additional 300 N forces were also applied in the normal direction of the aperture area.

Fig. 4.10 Ring used in the
collector by Al Asfar et al.
(2014)

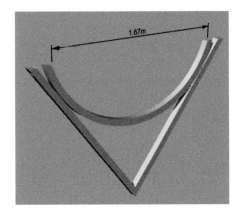

Fig. 4.11 Parabola
manufactured by Al Asfar et
al. (2014)

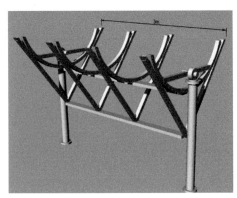

4.2.5 Al Asfar et al. (2014)

In this project [5], a low-cost PTC was constructed and tested. The total length of
the PTC is 6 m and the focal distance is 0.35 m.

The material used for the mirror is stainless steel, adopted in sheets of thickness
0.8 mm. It is exactly 304 BA (Bright Annealed) stainless steel which, once being
cold rolled, it is annealed in a controlled non-oxidant gaseous atmosphere to obtain
a highly reflective finish.

To sustain the aforementioned mirror, four triangular structures were made
(Fig. 4.10). They were linked by two bars, one of them in the base of the resul-
tant parabola and the other in the inferior corner of the triangle (Fig. 4.11). In order
to sustain it, three bars were used: two at the ends and the other between them. In
total, eight triangular structures, three solid subjection bars and four bars to link the
triangular and parabolic structures were used.

Once built the structure, several wind loads were performed. The load force was calculated as:

$$F_W = A_a p C_D \qquad (4.3)$$

where A_a is the aperture area, p is wind dynamic pressure and C_D is drag coefficient.

The results were obtained with a maximum value of 75 N/m due to the wind and a weight of the trough of 270 N/m. In addition, a mechanical analysis of the structure of finite elements was performed with ABAQUS software. With this analysis, it was possible to realize that the loading conditions were within the elastic limits, so it was possible to carry on with the project having verified that the external forces were supported without deforming the structure.

4.2.6 Coccia et al. (2014)

To realize the UNIVPM.01 prototype mold [6], reported in Fig. 4.12, the mold used by Kalogirou et al. [1] was used as a reference. The used material is light plastic instead of synthetic wood. To create the parabolic profile, nine pieces of 13 mm thick were cut (Fig. 4.13). A computer-controlled machine was used to guarantee the necessary accuracy, following the water-cut methodology. The separation between pieces was 247.9 mm, thus the total length, including the thickness of each plastic sheet, is 2.1 m. The width of the parabola is 1 m.

Once realized the mold, a 0.8 mm thick stainless steel sheet was laid on it, which, thanks to its own weight and two wooden beams, assumed a parabolic shape. To make the parabolic support structure, a 40 mm thick composite of extruded polystyrene (XEPS) was used mainly because of its lightness. This material was cut in small strips which were glued each other following a parabolic form. The free spaces were filled with epoxy resin. Two XEPS strips were replaced by two rectangular cross-section aluminum bars strategically located to obtain a stable structure and to link the concentrator to the support system. These bars were firmly attached to the other strips through four tubes positioned perpendicularly.

Fig. 4.12 Mold used for the manufacture of UNIVPM.01 (2014)

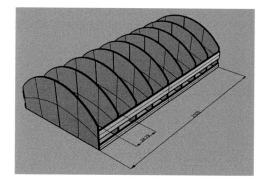

Fig. 4.13 A piece of light plastic used for the mold of UNIVPM.01 (2014)

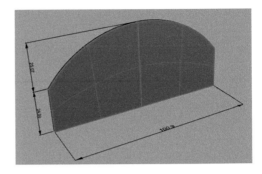

Fig. 4.14 UNIVPM.01 (2014) PTC design

In the end, a wax polish film was applied on all the necessary surfaces to achieve a proper finish. Then, the first and second fiberglass and resin layers were applied to the stainless steel parabolic surface. A normal mesh fiberglass was used instead of chopped strand fiberglass, which is cheaper but worse when thin layers are needed. While the compound was still liquid, all the XEPS strips and aluminum frame pieces were arranged in place. All remaining surfaces were covered with two layers of resin and fiberglass and it was carefully ensured that the resin penetrated well between the XEPS pieces. Once the resin was applied, the upper surface was covered with a plastic sheet and three straps were attached to it. Both the sheet and the straps were screwed down to the supports to ensure perfect adhesion of the new structure to the mold.

Once dry, the structure was removed from the mold. At this stage, a reflective layer made of aluminum was glued to the concave surface to create a parabolic reflective surface. This material was chosen because of its high reflection in the solar spectrum and resistance to atmospheric agents.

The final weight of the concentrator was roughly 12 kg (Fig. 4.14).

4.2.7 Kasaeian et al. (2014)

For their concentrator [7], the authors looked for lightness as well as resistance against wind loads. Consequently, wooden fixtures were employed and steel tubes connected the basal structure with the parabolic surface.

The main material was polyurethane, whose parabolic shape was obtained using a computer numerical control machine. This machine permitted to obtain a very precise parabolic profile adapted to the required features. The reflective material, made of 0.8 mm thick steel mirror, was obtained by a laser cutter.

The dimensions of the concentrator are relatively small being the parabola length equal to 2 m, the width equal to 0.7 m and 90° the rim angle. The focal distance was obtained with the following formula:

$$f = \frac{w}{2}\cot\phi + \frac{w^2}{16f} \tag{4.4}$$

where w is the aperture width and ϕ is the rim angle equal to 90°. The obtained value of 0.175 m was then introduced into the parabola Eq. (2.19) giving:

$$y = 1.428\,x^2 \tag{4.5}$$

4.3 Materials Used in PTC Concentrators

This section provides useful information about the materials generally used in the manufacture of PTC concentrators.

4.3.1 Aluminum

The most important aspect of a PTC concentrator is the reflectance of the material used as reflective foil. The average reflectance of this material just after installation (with no previous environmental effect) should be greater than 90% in the wavelength range between 250 and 2500 nm, which is generally considered the spectrum exploitable by solar collectors on Earth.

A number of solar reflective foils are made of aluminum. Aluminum is the third element in the Earth crust and the second behind the iron in level of consume. Pure aluminum itself has not very good mechanical properties, but if it is alloyed, it is quite improved. It has ease of forming and machining because of its low melting point (510 to 660 °C). What makes aluminum adequate for PTCs is its optical properties. In fact, its high light reflectance in a wide range of wavelengths and consequently low power absorption, allows it to reflect a relevant portion of the solar spectrum.

Commercially available solar reflective foils can often reach a reflectance in the solar spectrum of 97%, depending on the surface treatment used (mechanical, chemical or electrolytic polishing).

Reflective foils must adequately resist abrasion. Aluminum per se presents high resistance to corrosion, perfect to face variable environmental conditions. In fact, it reacts quickly exposed to aqueous environments or air because of its high affinity for oxygen and it forms a passive layer on the surface that protects against internal corrosion. Also, if this layer is damaged, it is easily regenerated. There are ways in which it may be more sensitive to corrosion, for example with lower aqueous environments at a pH of 4 or higher than 8.5.

The typical corrosion in metals is called pitting. Pitting corrosion causes small holes in the metal where the layer is not completely protective. If this corrosion pierces the metal, the reflecting properties of the foil could be compromised. Also, it should be noted that the passive layer, which protects the metal, has no influence for what concerns the reflection of light rays.

Generally, the producers of solar reflective foils carry out accelerated aging tests to prove that the material can stand in outdoor environments. Considering that to date there is no formula able to correlate aging tests with actual time in outdoor environments, care should been taken in choosing the proper product. Typical ageing tests usually include: abrasion, temperature, boiling water, UV-C, bend, salt and steam tests.

The mechanical properties of aluminum are adequate to be used in a solar collector not only to reflect the rays to the receiver, but they are also suitable to manufacture the structure of the concentrator. In some cases, it can be also used as material for the receiver, because of its good thermal conductivity.

4.3.2 Fiberglass

When the temperature increases up to the working point and glass associates with several oxides such as alumina or alkaline Earth metals, glass can be fibered obtaining the so-called fiberglass. This material has been known for long, but in 1938 it was known as fiberglass thanks to its inventor Russel Games. Its initial application was as an insulator material. Nonetheless, it currently has different applications, from hockey sticks to fabrication of splints in medicine. However, its main uses are insulation (acoustic, thermal and electrical) and reinforcement of polymeric products, which is known as fiberglass reinforced plastic and is commonly utilized due to its low-cost/rigidity-hardness ratio in comparison with other similar-characteristic fibers, such as carbon fiber.

There is a variety of glass with which the fiberglass can be made according to the different added chemical compounds (usually to improve fragility and hardness and decrease the working softening point). The basal compound of glass is silica (SiO_2), which is naturally found in sand. When it is in a polymeric form, it forms groups of SiO_2 forming covalent tetrahedrons with the silicon atom in the central position. The oxygen atoms are located at the corners.

The disadvantages of this material is the fragility and the high working temperature. Because of this, the fabrication price increases. To reduce the aforementioned problems, impurities such as Na_2CO_3 are usually introduced to reduce the working temperature and to offer major resistance.

The glass density is around $2500\,kg/m^3$, very similar to aluminum. The density varies considerably according to temperature, pressure and, above all, composition. For example, the more the concentration of calcium oxide (CaO), the more the density. On the other hand, the more aluminum (Al_2O_3), the less the density. The thermal conductivity is very low ($0.05\,W/m\,K$) because fiberglass, thanks to its chemical structure, is able to traps the air inside; because of this characterstic, it is usually used as a thermal insulator.

References

1. Kalogirou SA, Lloyd S, Ward J, Eleftheriou P (1994) Design and performance characteristics of a parabolic-trough solar-collector system. Renew Energy 5:384–386
2. Brooks M, Mills I, Harms T (2005) Design, construction and testing of a parabolic trough solar collector for a developing-country application. Proceedings of the ISES Solar World Congress, Orlando, 605:6–12
3. Valan Arasu A, Sornakumar T (2007) Design, manufacture and testing of fiberglass reinforced parabola trough for parabolic trough solar collectors. Sol Energy 81:1273–1279
4. Jaramillo OA, Venegas-Reyes E, Aguilar JO, Castrejn-Garca R, Sosa-Montemayor F (2013) Parabolic trough concentrators for low enthalpy processes. Renew Energy 60:529–539
5. Al Asfar J, Ayadi O, Al Salaymeh A (2014) Design and performance assessment of a parabolic trough collector. Jordan J Mech Ind Eng 8:1–5
6. Coccia G, Di Nicola G, Sotte M (2014) Design, manufacture, and test of a prototype for a parabolic trough collector for industrial process heat. Renew Energy 74:727–736
7. Kasaeian A, Daviran S, Azarian RD, Rashidi A (2014) Performance evaluation and nanofluid using capability study of a solar parabolic trough collector. Energy Convers Manag 89:368–375

Chapter 5
Receiver

Abstract Receivers are the parts of a PTC where the solar radiation is concentrated and collected. PTC receivers are tubes of various diameters located in the focal line of the concentrator and can be distinguished into two parts: the absorber, i.e. the metallic tube in which the heat transfer fluid (HTF) flows, and the cover, an external tube generally made of glass, adopted to reduce convective and radiative thermal losses. The absorber collects the solar radiation reflected and focused by the concentrator, and it transfers the absorbed heat to the HTF by conduction and then by convection. Due to their function, receivers should have high thermal conductivity and present high absorptance to the solar spectrum. The latter condition is generally realized by painting the outer surface of the receiver with special black coatings. This chapter presents an overview of the receivers used in PTC prototypes, providing some characteristics about the adopted materials. The performance of the prototypes are also discussed. Finally, PTC protoypes using nanofluids, novel promising HTFs, are presented.

Keywords Selective coating · Vacuum · Nanofluid · Convective heat transfer · Thermal conductivity

5.1 Introduction to PTC Receivers

Along with the concentrator, the second important issue in the structure of a PTC lies in the device where solar rays are directed by the concentrator itself (Fig. 5.1). This element is named receiver and is located on the focal line of the parabolic concentrator. It is generally composed by a metallic tube covered by a black layer (a paint or varnish) and can be or not shielded with another transparent tube which is useful to reduce convective and radiative heat losses.

A typical PTC receiver can be seen in Fig. 5.2. The heat transfer fluid (HTF) travels inside the metallic tube. The HTFs generally used are water and silicon oils; recently, research is focusing its attention to nanofluids, which will be discussed in the following.

© The Author(s) 2016
G. Coccia et al., *Parabolic Trough Collector Prototypes for Low-Temperature Process Heat*, SpringerBriefs in Applied Sciences and Technology,
DOI 10.1007/978-3-319-27084-5_5

Fig. 5.1 Cross-section of a
PTC concentrator and
receiver (cover and absorber
tube)

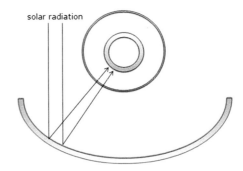

Fig. 5.2 Components of a
PTC receiver

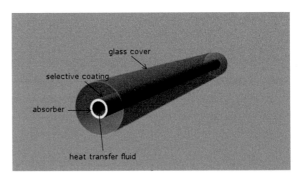

In order to choose the proper HTF, it is necessary to take into account the climatological extremes, the desired thermal efficiency and the economic costs. In the projects treated in this book, two different fluids have been utilized: mineral oil and demineralized water. Several tables have been provided indicating the cinematic viscosity, the thermal stability and the thermal conductivity of such fluids. In particular:

- The cinematic viscosity is the quotient between the absolute viscosity and the fluid density. The viscosity represents the internal resistance of a fluid to the flow. The cinematic viscosity is expressed in centistoke (cst) or m^2/s.
- The thermal stability is defined as the point in which all the portions of the fluid do not differ by more than $3\,°C$. In solar thermal applications, the aim is to reach the desired temperature in the minimum time. Its unit is seconds (s).
- The thermal conductivity is the property of a material to conduct heat throughout its molecules in a conductive process. Its unit is $W/(m\ K)$.

One of the essential characteristics of the HTFs used in solar thermal applications is to reach the maximum performance in homogeneity and to avoid any particle in suspension. The presence of particles can seriously damage the circuit valves and other hydraulic components which transport the fluid from a component to another one.

When comparing water with a nanofluid, the first aspect to address is the cost. Water is the cheapest fluid. However, nanofluids recently adopted in PTC prototypes seem to be more efficient. For example, in the project of Kasaeian et al. [6], the

Table 5.1 Characteristics of the different receivers

Description	[1] Shield.	[1] Unsh.	[2]	[3] PTC90	[3] PTC45	[4]	[5]	[6] Copper	[6] Steel
Inner abs. diameter (mm)	–	–	–	–	–	25	25	26	16.2
Outer abs. diameter (mm)	28.6	28.6	12.8	25.4	25.4	32	30	28	18
Inner cover diameter (mm)	39.4	–	–	–	–	–	46	58	58
Outer cover diameter (mm)	44	–	22.6	–	–	–	48	60	60
Length (m)	5.380	5.380	1.25	2.50	2.50	6	2.1	1.8	1.8
Absorber area (m^2)	0.48	0.48	0.09	0.20	0.20	0.57	0.20	0.16	0.10
Cover transmittance	0.92	–	0.9	–	–	–	0.93	–	–
Absorber absorptance	0.88	0.88	0.9	0.90	0.90	0.85	0.95	0.98	0.91
$F_R \eta_0$	0.538	0.552	0.69	0.613	0.351	–	0.658	0.6730	–
$-(F_R U_L)/C$	1.059	2.009	0.39	2.302	2.117	–	0.683	0.2243	–

efficiency increases by 7% as they use a mineral oil in which carbon nanotubes are suspended.

A summary of the characteristics of the proposed receivers, including the thermal efficiency coefficients (Sect. 3.4.2) of the whole PTC prototypes, are reported in Table 5.1.

5.1.1 Brooks et al. (2005)

In this project, two receivers were designed: one with a cover glass and the other one unshielded. Both of them have the same material and dimensions. The material used for the absorbers was copper and the diameter of the tubes was 28.6 mm. With the intention of increasing the absorptance and decreasing the emittance, both surfaces were painted with a selective coating.

In order to elaborate the glass cover, borosilicate was utilized (see Table 5.2). This material is a specific type of glass that contains approximately 80% of silicon oxides and 15% of boron oxides; the rest was aluminum, sodium and potassium oxides. This glass has great applications in both chemistry and engineering. Its high chemical resistance and low thermal expansion coefficient make this material appropriate for

Table 5.2 Borosilicate properties

Property	Value
Coeff. of thermal expansion (K^{-1})	3.3×10^{-6}
Density (kg/m^3)	2200
Refractive index	1.47

realizing test tubes and glass laboratory devices. In this case, where it is used as an external tube and high temperatures are reached, a low dilatation coefficient is desirable. It was modeled from a chemical glass blower process.

In addition, a possible expansion of the inner absorber was taken into account and two high-temperature-resistant O-rings were used to seal the ends of the annulus. The HTF was water with a volumetric flow of $0.3 \, m^3/h$ to ensure a turbulent flow inside the absorber.

In order to assess the performance of the PTC, the ANSI/ASHRAE Standard 93-1986 [7] was followed. The time constant for the shielded receiver was 28.6 s and for the unshielded arrangement was 30.5 s. The coefficients of the experimental thermal efficiency equation are reported in Table 5.1.

5.1.2 Valan Arasu and Sornakumar (2007)

The design of this receiver is based on a recipient tube made of copper with 12.8 mm in diameter. It is covered with a heat-resistant-black paint to improve the absorption of the solar radiation.

The tube is located on the focal line at 0.196 m from the concentrator. Furthermore, this line serves as axis of rotation in the tracking system; the prototype rotates around the horizontal north/south axis. The movement is controlled by a low-velocity DC motor of 12 V.

There is a low-iron glass tube which encloses the copper tube. Although the maximum temperature in the storage tank is 75 °C, the glass cover is used to increase the efficiency. The tube has an external diameter of 22.6 mm. In the annular space between the cover and the absorber, two rubber corks were added at both ends of the receiver to achieve an air-light enclosure.

Taking into account the declared optical properties, the optical performance is 0.694. The tests were performed according to the ANSI/ASHRAE Standard 93-1986 [7] and using water as HTF.

5.1.3 Jaramillo et al. (2013)

As mentioned in Chap. 4, for this project two concentrators were manufactured with a rim angle of 45° (PTC45) and 90° (PTC90), even though the used receiver remains the same.

In this case, the receiver is a tube without the glass cover in order to reduce the potential costs during both manufacturing and transporting. In fact, there are several cases where the tubes are not produced in the local country and the transport is highly expensive. The temperatures at which the tubes are exposed (70–110 °C) also influenced in the decision of not using the glass cover. For temperatures near 100 °C, the fact of having a glass cover does not increase significantly the PTC efficiency, so then the cost/efficiency ratio is not good.

The external diameter of the absorber was determined by using a ray tracing method technique. This method consists in carrying out a simulation of the sun rays direction, focusing them on a single line, i.e. the focal line. On this line, a circle with a 25.4 mm diameter is generated, which is used as the external diameter. To reduce costs, the absorber was realized in aluminum.

The HTF used during the test is water. The flow in both collectors was 4 L/min. With the intention of obtaining the time constants, Eq. (3.5) was used; the obtained constants were 35.5 s for PTC90 and 31.5 s for PTC45.

Comparing the thermal efficiency coefficients (Table 5.1), it is possible to observe that in the PTC90 the efficiency at $T_{f,in} = T_{amb}$ is almost 1.7 times higher than PTC45. The difference between each other is considerable, taking into account that the same material is utilized to make the concentrator and the receiver, and the conditions referring the sun rays are also the same. It can be concluded that the rim angle, the concentration ratio and the aperture area have a great influence.

5.1.4 Al Asfar et al. (2014)

In this work, there is no specification regarding the metal used to realize the receiver. It is supposed that it is steel because this material was used in every other part of the prototype. However, the absorber tube was covered with a black painting having an absorptivity in the solar spectrum up to 85%. The receiver is provided with a glass cover.

The external diameter is equal to 32 mm and the tube is 3.5 mm thick. The authors used pressurized water as HTF, with a maximum outlet fluid temperature of 127 °C.

The time constant was obtained through Eq. (3.5) and it is equal to 4.25 min. The mean of three different volumes of water (52.2, 69.2 and 90 L/h) was used to calculate the time constant.

5.1.5 Coccia et al. (2014)

The design of this PTC (UNIVPM.01) allows to easily replace the receiver under test with other tubes of different characteristics. In this case, lightness, ductility and cost were fundamental, so aluminum seemed to be perfect for the purpose. The surface of the absorber was painted with a black-high-temperature-resistant paint. In addition, the temperature reached during the experiments was 85 °C, so there were no problems to work with this material.

The external diameter was 30 mm and the thickness 2.5 mm. Even though the working temperature was not high, in this project a low-iron glass was used to increase the efficiency and to minimize the heat losses. The glass has an external diameter of 48 mm and a thickness of 1 mm. According to this measurements, the annulus between the absorber and the glass cover is 8 mm. The focal line was placed at 0.25 mm and the total length of the receiver was 2.1 m. This length is the same of the concentrator.

One of the problems associated with the annulus space is condensation. To overcome this problem, the three teflon rings used to separate the absorber and the cover were drilled.

The HTF was demineralized water. Although the ANSI/ASHRAE Standard 93-2010 recommends a value of 0.037 kg/s for 1.85 m^2 of aperture area, the mass flow rate was set to 0.045 kg/s. This value was used in order to obtain a turbulent flow regime in the absorber. Other characteristics of the receiver adopted in this project, including the performance of the whole collector, are reported in Table 5.1.

5.1.6   Kasaeian et al. (2014)

In this project, tests for the optical and thermal performances were carried out using four different receivers. The absorber tubes were realized of steel or copper. With the latter, three absorbers were made, using a black chrome coating. On the other hand, the unique steel design was painted with a resistant matte black paint.

Both the inlet and the outlet of the tube were sealed with two incombustible teflon couplings with the intention of avoiding welding. A valve was installed to evacuate or refill the HTF in case of troubles. This valve also served for replacing the absorber tubes. In order to avoid that a too high tension due to high temperatures was transmitted to the glass tube, O-rings were introduced around the grooves of the teflon couplings.

Mineral oil was used as the HTF (see Table 5.3), which has more viscosity than water. This fluid was controlled by two taps, one to control the main line and the other one to control the by-pass line. The fluid is bombed from a reservoir tank to the tube through a helical duct made of PVC and covered by glass wool, to reduce heat losses. The reservoir tank is placed in a constant-temperature bath to stabilize

Table 5.3 Mineral oil properties	Description	Value
	Cinematic viscosity (m^2/s)	32×10^{-6}
	Thermal conductivity (W/(m K))	0.128

the inlet temperature monitored with thermometers. At the end of the tube, a flow meter and a sight glass were also included to control the fluid conditions.

Into the mineral oil, multi-walled carbon nanotubes (MCNTs) were introduced with the intention of increasing the thermo-physical properties of the HTF. Different processes were performed to avoid the agglomeration along the fluid and ensure a proper homogenization and separation between the particles and the fluid. It was confirmed that introducing 0.2–0.3% of MCNTs, the process efficiency increases by 4–7%.

The overall results were analyzed with four kind of absorber tubes:

1. a black painted vacuumed steel tube;
2. a copper bare tube with black chrome coating;
3. a glass enveloped non-evacuated copper tube with black chrome coating;
4. a vacuumed copper tube with black chrome coating.

The first operation carried out by the authors was to calculate the stabilization time constant (using Eq. (3.5)) of the system for the different cases. The constant are, respectively: 11.37, 8.53, 8.48 and 9.98 s. The arrangement where the temperature was stabilized quicker is in the copper absorber tube with glass envelope; and the slowest is the steel tube, due to the waste of heat with radiation.

5.2 Materials Used in PTC Receivers

The absorber is a crucial element of a PTC since its inner surface is in contact with the HTF and the outer surface collects the solar radiation. Thus, its thermal and optical properties must be adequate.

Materials generally used for PTC absorbers are:

- copper (as in [1, 2, 6]);
- aluminum (as in [3, 5]);
- stainless steel ([6]).

The thermo-physical properties of such materials are quite different, as reported in Table 5.4. Among these, one of the most important is thermal conductivity. In fact, for aluminum and copper it is 4–5 times and 7–8 times bigger than that of steel, respectively. Something similar happens with the thermal diffusivity, with the same proportions.

5.2.1 Copper

Copper is, along with aluminum and iron, the most consumed metal in the world. It is soft (3 in Mohs scale) but nevertheless it has a high wear resistance. Copper is

Table 5.4 Thermal properties of the materials used in PTC absorbers

Materials	ρ (kg/m^3)	c_p (J/(kg K))	λ (W/(m K))	Diffusivity (m^2/s)
Steel	7850	460	47–50	13–16
Aluminum	2700	909	209–232	85–95
Copper	8900	389	372–385	107–112

used specially in the electrical lines, as it is ductile and malleable. It is a very good electrical driver, in fact, its conductivity is the standard reference for this magnitude.

Its mechanical properties are low and, in order to improve them, other elements are usually used to form alloys. The most common elements combined with copper are zinc, tin and nickel. When its alloyed with zinc, the resulting material is called brass, which usually carries a percentage from 5 to 55%, although a too high percentage can give fragility. When it is alloyed with tin, the result is bronze. Here copper takes a percentage from 2 to 22%.

Apart from the fact that its thermal conductivity is high, another quality that makes copper suitable for use in solar absorbers is its high resistance to corrosion. Its positive value potential is 0.35, which provides a big stability and because of this it is considered a noble metal, together with silver and gold. This stability comes for having a positive potential, for what it tends to admit electrons. Also, in presence of oxygen, copper creates a thin protective and insulating layer of oxide on the surface.

5.2.2 Steel

Steel is an iron alloy with a maximum percentage in carbon equal to 1.7%. There are different types of steel depending on the percentage in carbon and the addition of different alloying. Other elements which can be added are: manganese which increments its hardness and wear resistance; nickel, that improves toughness, tensile strength and resistance to corrosion; and other components such as molybdenum, vanadium, silicon, etc. which remain on the line to improve its mechanical properties.

Stainless steels are usually adopted in PTC absorbers. These steels are characterized by being composed of at least 12% of chromium. Chromium is closely related to oxygen and reacts with it producing a chromium oxide film that makes waterproof its inner steel structure and prevents oxygen from passing, thus avoiding a reaction with the steel. Another positive feature of this film (passive layer) is that it is self-healing in the presence of oxygen. In addition, if it is mechanically or physically damaged, as it can be due to any touch or scratch on the receptor pipe surface, the protective layer is created again.

Among stainless steels, there are also different types according to the percentages of carbon, chromium and nickel used. An example is the austenitic steel which contains high level contents of nickel and chromium. Its main feature is that it is not magnetic. One of the most common contains the 18% of chromium and the 8% of steel and it is mainly used to make kitchen utensils.

The ferritic steel is a type of stainless steel with 16–18% of chromium, less than 0.1% of carbon and less than 8% of nickel. If the percentage of carbon is increased with the appropriate heat treatment, martensitic steel can be obtained. What makes it different is its high hardness. There is also the so-called duplex stainless steel which is a mix between austenitic and ferritic types.

5.2.3   Selective Coatings

In order to increase the thermal efficiency of a PTC and collect the maximum amount of solar radiation, the receiver is painted in black and also covered with a thin selective coating. These layers are introduced for getting the absorber closer to behave as a blackbody. It should have high absorbance (around 0.90–0.95) in the wavelength range of the solar spectrum and low emissivity (less than 0.2) in the infrared region at the working temperature of the HTF.

The color black helps in increasing the absorbance, but the emissivity would be too high if only this is applied. In fact, if the fluid reaches high temperatures, the tube acts as an emitting radiator, losing energy. Therefore, the more the absorbance and the lower the emissivity, the better the performance.

There is a correlation between both parameters, the selectivity. This is the quotient between the absorbance and emissivity. Hence, when selecting the selective coating, the selectivity should be as higher as possible. Different physical and chemical parameters related to the emissivity and absorbance can led an increment and decrement of this factor. For example, the more polished the external layer of the receiver, the less the emissivity and the more the selectivity. The discoloration of the selective layer pigment is an example of the decrement of selectivity of the tube. This favors the absorbance decrement and, therefore, the selectivity also decreases.

The majority of the carried-out treatments consists in using metallic oxides on a metallic substrate. The substrate is in charge of reducing the emissivity and the oxides increase the absorbance. The metal usually has a ceramic-matrix layer where the metallic particles are embedded after depositing them by means of cathodic pulverization. The obtained material is called cermet (ceramic + metallic). This layer captures the visible radiation. There are different types of cermets based on several metals, such as molybdenum, cobalt, nickel with a pigment anodic of Al_2O_3 produced by electrochemical treatment of an aluminum sheet, or the most used black chrome ($Cr-Cr_2O_3$).

In order to apply this selective layer, the most used technique is electro-positioning because of its simplicity and low cost. Other methods exist such as CVD (Chemical Vapor Deposition) or PVD (Physical Vapor Deposition), but they are more expensive. The unique inconvenient of this covering is that the maximum working temperature is up to 400 °C. With higher temperatures, the layer starts degrading. Research is working to increase the maximum temperature without degradation.

5.3   Nanofluids

The PTC technology was largely subsidized and developed during the last decade [8–10]. Today, the most urgent demand consists of increasing the thermal efficiency of these systems: this is particularly true for low-enthalpy PTCs. One possible solution to improve the thermal efficiency of such systems could lie in the use of nanofluids as heat transfer fluids. In fact, it is reasonable to expect an increase in the thermal efficiency of low-enthalpy PTCs when the heat transfer base fluid is substituted with a nanofluid of appropriate concentration of nanoparticles.

In their paper of 1995 [11], Choi and Eastman introduced the term nanofluid to describe a new class of heat transfer fluids obtained through different processes, such as chemical synthesis. A nanofluid is a dispersion of solid particles with nanometric dimension in a base fluid (e.g., water). Different types of nanoparticles were studied in literature: oxide ceramics, nitride ceramics, carbide ceramics, metals, semiconductors, carbon nanotubes, carbon nanohorns, and composite materials. Several works report that the thermal conductivity and the convective heat transfer coefficient of nanofluids exhibit significant enhancements with respect of the base fluid if the suspension is stable, even with a reduced quantity of nanoparticles.

One of the most important thermo-physical properties in PTCs is the thermal conductivity. With respect of the base fluid, a nanofluid presents a higher thermal conductivity because the solid nanoparticles have a higher conductivity than liquids. It has been proved that even having a low concentration of nanoparticles, the conductivity increases.

The use of nanofluids is progressing as heat transfer fluids. Nanofluids with nanoparticles smaller than 100 nm have been used in heat exchangers to increase the efficiency and the heat exchange.

Table 5.5 provides the thermo-physical properties of the nanoparticles used in the projects described in the following sections.

5.3.1   De Risi et al. (2013)

In this project, the results of a mathematical modeling using a nanofluid with nanoparticles of CuO and Ni having a concentration volume of 0.3% were studied [12]. The

Table 5.5 Thermo-physical properties of nanoparticles

Nanoparticle	ρ (kg/m^3)	c_p (J/(kg K))	λ (W/(m K))	Diffusivity (m^2/s)
Al	2700	909	209–232	85–95
Ni	8800	460	52.3	12.92
Au	19330	234	308.2	122.65
Ag	10500	234	418	170.13

addition of CuO was analyzed and it was proved that a complete absorption of the solar spectrum inside the receiver was achieved.

In a later stage, Ni was also added. In terms of economy, it is not advisable, because Ni is about 20 times more expensive than metal-oxide nanoparticles. The optimal ratio was about 0.25% of CuO and a 0.05% of Ni. It was found that there is a correlation between the concentration of nanoparticles and the PTC thermal efficiency. The maximum thermal efficiency, equal to 0.625, was obtained with the aforementioned proportion of nanoparticles at an outlet fluid temperature of 650 °C.

5.3.2 Sokhansefat et al. (2014)

In the work of Sokhansefat et al. [13], synthetic oil was used as a base fluid and Al_2O_3 as solid nanoparticles. The study was conduced at four different volume concentrations of nanoparticles: pure synthetic oil and then 1, 3 and 5%. Work temperatures were 300, 400 and 500 K.

Table 5.6 shows the variation of the thermo-physical properties as a function of the volume concentration at room temperature. With the exception of the heat capacity, it can be observed that each property increases with the volume concentration.

In terms of density, it can be observed how it increases as more particles are added. From $936.5\,kg/m^3$ in pure oil to $1082.1\,kg/m^3$ in 5% concentrate oil, it increases of 15% exactly. With the conductivity, something similar occurs, with an increment of about 20%. Viscosity has a parabolic trend, instead.

The conclusions of this study are that when the nanofluid is used, the heat transfer increases, being able to reduce the transfer area (saving money in terms of material). Furthermore, it can be proved that the higher the concentration is, the higher yield has the PTC.

Table 5.6 Thermo-physical properties at different volume concentration in the work of Sokhansefat et al.

Concentration	ρ (kg/m^3)	ν (mPa s)	λ (W/(m K))	c_p (J/(kg K))
0	936.5	6.68	0.134	1620
0.01	965.6	6.84	0.139	1612
0.03	1023.9	7.18	0.150	1595
0.05	1082.1	7.518	0.161	1578

5.3.3   Waghole et al. (2014)

Waghole et al. [14] carried out a study in 2013 with water and silver nanoparticles. The maximum volume concentration of nanoparticles was 0.1%. Tests were made in a PTC with a copper receiver 1.5 m long and an inner diameter of 20 mm.

The flow regime was set to a number of Reynolds between 500 and 6000, and the following parameters where studied: the Nusselt number, the friction factor and the thermal efficiency. The study concluded comparing the nanofluid with pure water, and the results were as follows:

- the Nusselt number increased in order of 1.25–2.10 times;
- the friction factor increased from 1 to 1.75 times;
- the thermal efficiency was also higher, between 135–205%.

References

1. Brooks M, Mills I, Harms T (2005) Design, construction and testing of a parabolic trough solar collector for a developing-country application. Proceedings of the ISES Solar World Congress, Orlando, FL 605:6–12
2. Valan Arasu A, Sornakumar T (2007) Design, manufacture and testing of fiberglass reinforced parabola trough for parabolic trough solar collectors. Sol Energy 81:1273–1279
3. Jaramillo OA, Venegas-Reyes E, Aguilar JO, Castren-Garca R, Sosa-Montemayor F (2013) Parabolic trough concentrators for low enthalpy processes. Renew Energy 60:529–539
4. Al Asfar J, Ayadi O, Al Salaymeh A (2014) Design and performance assessment of a parabolic trough collector. Jordan J Mech Ind Eng 8:1–5
5. Coccia G, Di Nicola G, Sotte M (2014) Design, manufacture, and test of a prototype for a parabolic trough collector for industrial process heat. Renew Energy 74:727–736
6. Kasaeian A, Daviran S, Azarian RD, Rashidi A (2014) Performance evaluation and nanofluid using capability study of a solar parabolic trough collector. Energy Convers Manag 89:368–375
7. ANSI/ASHRAE Standard 93–1986 (1986) Methods of testing to determine the thermal performance of solar collectors
8. Price H, Kearney D, Cohen G, Mahoney R (2002) Advances in parabolic trough solar power technology. J Sol Energy Eng-T ASME 124:109–125
9. Kalogirou SA (2003) The potential of solar industrial process heat applications. Appl Energy 76:337–361
10. Fernandez-Garcia A, Zarza E, Valenzuela L, Perez M (2010) Parabolic-trough solar collectors and their applications. Renew Sustain Energy Rev 14:1695–1721
11. Choi SUS, Eastman JA (1995) Enhancing thermal conductivity of fluids with nanoparticles. ASME Publications Fed 231:99–106
12. De Risi A, Milanese M, Laforgia D (2013) Modeling and optimization of transparent parabolic trough collector based on gas-phase nanofluids. Renew Energy 58:134–139
13. Sokhansefat T, Kasaeian AB, Kowsary F (2014) Heat transfer enhancement in parabolic trough collector tube using Al_2O_3/synthetic oil nanofluid. Renew Sustain Energy Rev 33:636–644
14. Waghole DR, Warkhedkar RM, Kulkarni VS, Shrivastva RK (2014) Experimental investigations on heat transfer and friction factor of silver nanofluid in absorber/receiver of parabolic trough collector with twisted tape inserts. Energy Procedia 45:558–567

Printed in the United States
By Bookmasters